D0875492

The Feeling of Life Itself

The Feeling of Life Itself

Why Consciousness Is Widespread but Can't Be Computed

Christof Koch

The MIT Press
Cambridge, Massachusetts
London, England

This book was set in Stone Serif and Stone Sans by Westchester Publishing Services. Printed and bound in the United States of America.

Library of Congress Cataloging-in-Publication Data is available.

ISBN: 978-0-262-04281-9

10 9 8 7 6 5 4 3 2 1

To Teresa

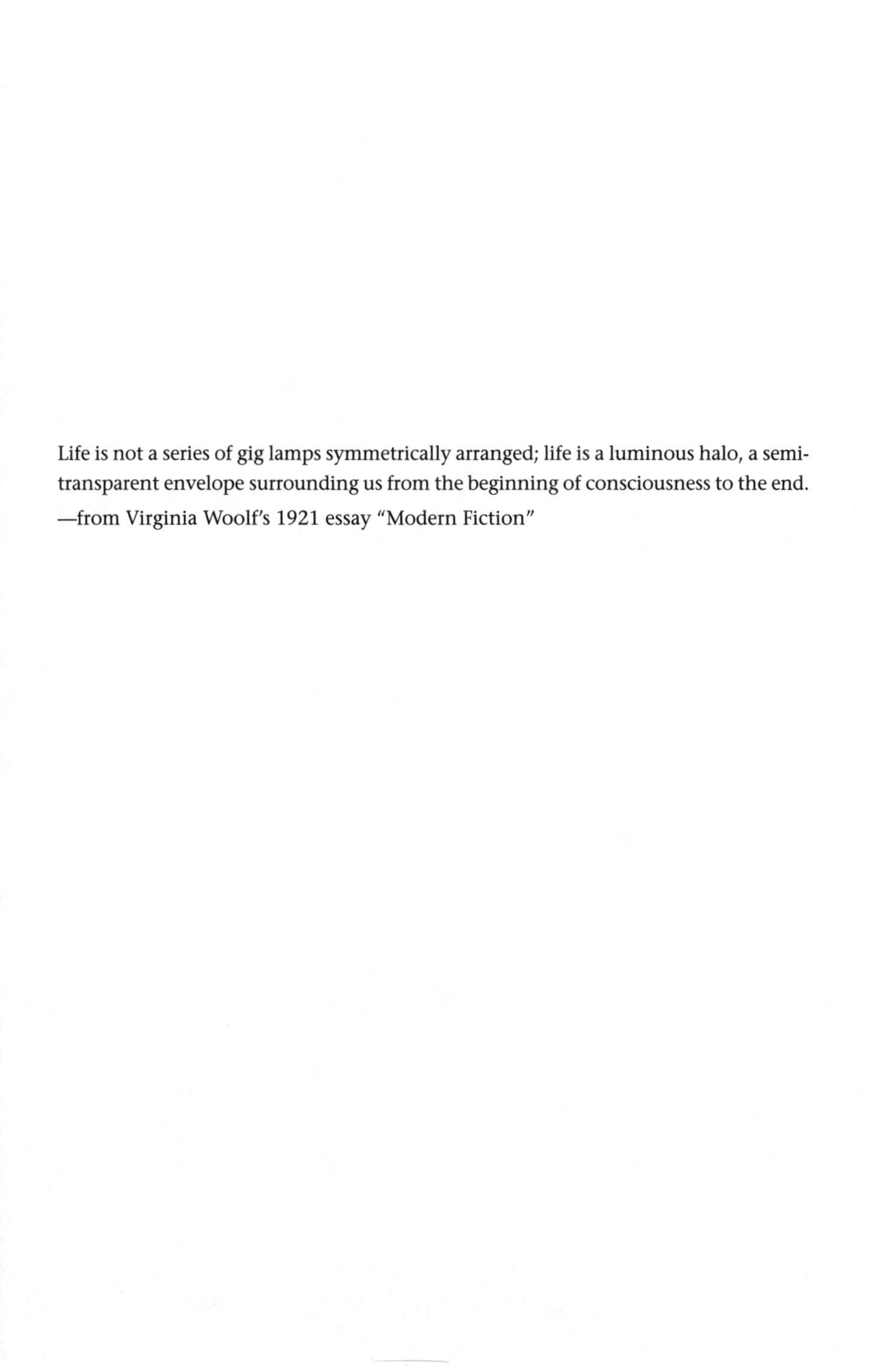

Life is not a series of gig lamps symmetrically arranged; life is a luminous halo, a semi-transparent envelope surrounding us from the beginning of consciousness to the end.
—from Virginia Woolf's 1921 essay "Modern Fiction"

Contents

Preface: Consciousness Redux xi
Acknowledgments xvii

1 **What Is Consciousness?** 1
2 **Who Is Conscious?** 11
3 **Animal Consciousness** 25
4 **Consciousness and the Rest** 33
5 **Consciousness and the Brain** 39
6 **Tracking the Footprints of Consciousness** 53
7 **Why We Need a Theory of Consciousness** 71
8 **Of Wholes** 79
9 **Tools to Measure Consciousness** 93
10 **The Über-Mind and Pure Consciousness** 105
11 **Does Consciousness Have a Function?** 119
12 **Consciousness and Computationalism** 129
13 **Why Computers Can't Experience** 141
14 **Is Consciousness Everywhere?** 155

Coda: Why This Matters 169
Notes 175
References 211
Index 243

Preface: Consciousness Redux

To bring home the centrality of consciousness to life, consider a devil's bargain in which you gain unlimited wealth at the expense of your conscious experiences. You get all the money you want but must relinquish all subjective feeling, turning into a zombie. From the outside, everything appears normal—you speak, act, dispose of your vast riches, engage in a vigorous social life, and so on. Yet your inner life is gone; no more seeing, hearing, smelling, loving, hating, suffering, remembering, thinking, planning, imagining, dreaming, regretting, wanting, hoping, dreading. From your point of view, you might as well be dead, for it would feel the same—like nothing.

How experience comes into the world has been an abiding mystery since the earliest days of recorded thought. Aristotle warned his readers more than two thousand years ago that "to attain any assured knowledge about the soul is one of the most difficult things in the world." Known as the *mind–body problem*, this puzzle has occupied philosophers and scholars throughout the ages. Your subjective experience appears radically different from the physical stuff that makes up your brain. The foundational equations of physics, the periodic table of chemical elements, the endless ATGC chatter of your genes—none of these express anything about consciousness. Yet you awaken each morning to a world in which you see, hear, feel, and think. Experience is the only way you know about the world.

How does the mental relate to the physical? Most assume that the mental emerged from the physical when the physical became sufficiently complex. That is, in the eons before big brains like ours evolved on this planet, the mental did not exist. Yet are we really to believe that until that point in time (in a memorable phrase of the physicist Erwin Schrödinger), the world "remained a play before empty benches, not existing for anybody, thus quite properly speaking not existing"? Alternatively, perhaps the mental was always present, allied with the physical, but not in a form readily recognizable? Perhaps consciousness predates the arrivals of big brains? This is the less-traveled road that I will take here.

When did *your* consciousness begin? Was your first experience one of confusion and chaos as the birthing process ejected you into a harsh world, blinding you with bright lights, overwhelming sounds, and a desperate need for oxygen? Or perhaps even earlier, of warmth and security in your mother's womb?

How will your stream of experience end? Snuffed out abruptly like a candle, or fading gradually? Can your mind, chained to a dying animal, encounter the numinous in a near-death experience? Can advanced technology come to the rescue and transition your mind to an engineered medium, a ghost in a new shell, by uploading your mind to the Cloud?

Do apes, monkeys, and other mammals hear the sounds and see the sights of life? Are dogs just machines, as René Descartes famously argued, or do they experience the world in a riot of redolent smells?

Then there is the urgent question of the day—can computers experience anything? Can digital code feel like something? Dramatic progress in machine learning has crossed a threshold, and human-level artificial intelligence may come into the world within the lifetime of many readers. Will these AIs have human-level consciousness to match their human-level intelligence?

In this book, I will show how these questions, formerly the sole province of philosophers, novelists, and moviemakers, are now being addressed by scientists. Powered by advanced instrumentation that peers deep into the brain, the science of consciousness has seen dramatic progress over the past decade. Psychologists have dissected out which cognitive operations underpin any one conscious perception. Much cognition occurs outside the limelight of consciousness. Science is bringing light to these dark passages where strange, forgotten things live in the shadows.

I dedicate two chapters to tracking the footprints of consciousness within its principle organ, the nervous system. Surprisingly, many brain regions do not contribute meaningfully to experience. This is true for the cerebellum, despite having more than four times more nerve cells than neocortex. Even in neocortical tissue, the most complex piece of highly excitable matter in the known cosmos, some sectors have a much more intimate relationship to experience than others.

In the fullness of time, the quest for the neural footprints of consciousness will pursue its prey to its lair somewhere in the thicket of the nervous system. Sooner or later, scientists will know which assemblies of nerve cells, expressing which proteins, and active in some mode, house any one experience. This discovery will be a high water mark for science. It will be enormously beneficial to neurological and psychiatric patients.

Yet knowing the neural correlates of consciousness does not answer the more fundamental question: Why these neurons and not those? Why this vibration and not that one? Identifying some form of physical activity as generating a feeling is laudable progress. But ultimately we want to know why this mechanism goes hand in hand with experience. What is it about the biophysics of the brain, but not, say, the liver—another complex biological organ—that evokes the ephemeral feelings of life?

What we need is a quantitative theory that starts with experience and proceeds from there to the brain. A theory that infers and predicts where experience can be found. To me, the most exciting development over the last decade has been the genesis of such a theory, a first in the history of thought. Integrated information theory considers the parts and their interactions that make up a whole, whether evolved or engineered, and derives, via a well-specified calculus, the quantity and the quality of the experience of this whole. The beating heart of *The Feeling of Life Itself* comprises two chapters outlining the theory and how it defines any one conscious experience in terms of intrinsic causal powers.

From the austerity of these abstract considerations, I dive into messy clinical practice. I describe how the theory has been used to build a tool to detect the presence and absence of consciousness in unresponsive patients. I next discuss some of the theory's counterintuitive predictions. If the brain is cut in the right place, its unitary mind splits into two minds that coexist within a single skull. Conversely, if the brains of two people are directly connected via a futuristic *brain-bridging* technology, their distinct minds could fuse into a single mind at the cost of their individual minds, which will be extinguished. The theory predicts that consciousness without any content, known as *pure experience* within certain meditative practices, can be achieved in a near-silent cortex.

After considering why consciousness evolved, *The Feeling of Life Itself* turns to computers. The basic tenet of today's dominant faith, its zeitgeist, is that digital, programmable computers can, in the fullness of time, simulate anything, including human-level intelligence and consciousness. Computer experience is just a clever hack away.

According to integrated information theory, nothing could be further from the truth. Experience does not arise out of computation. Despite the near-religious belief of the digerati in Silicon Valley, there will not be a Soul 2.0 running in the Cloud. While appropriately programmed algorithms can recognize images, play Go, speak to us and drive a car, they will never be conscious. Even a perfect software model of the human brain will not experience anything, because it lacks the intrinsic causal powers of the brain. It will act and speak intelligently. It will claim to have experiences, but that will be make-believe—fake consciousness. No one is home. Intelligence without experience.

Consciousness belongs to the natural realm. Just like mass and charge, it has causal powers. To create human-level consciousness in a machine, the intrinsic causal powers of the human brain must be instantiated at the level of the metal, the transistors and wiring making up its hardware. I will show that the intrinsic causal powers of contemporary computers is puny compared to those of brains. Thus, artificial consciousness calls either for computer architectures radically different from those of today's machines or for a merging of neural and silicon circuits as envisioned by transhumanists.

In the concluding chapter, I survey nature's wide circuit. Because of the vast complexity of brains of so-called simple animals, IIT implies experience in parrots, ravens, octopuses, and bees. As nervous systems devolve into the primitive nerve net of jellyfish, their associated experience will lessen. But single-cell microorganisms contain untamed molecular complexity inside their cellular envelope, so they too are likely to feel an itsy-bitsy bit like something.

Integrated information theory has captured the imagination of philosophers, scientists, and clinicians as it opens myriads of doors to experimentation and because of its promise to illuminate those aspects of reality that have been, until now, beyond the pale of empirical investigations.

Any entrepreneur launching a new company against overwhelming odds must have a healthy amount of self-delusion. This is essential to remain motivated to work crazy hours year after year. Accordingly, I wrote this book assuming the theory is true and discard the scholar's cautionary habit of prefacing every statement with "under certain conditions." I do note current controversies and cite extensive and up-to-date literature in the notes. Of course, at the end of the day, nature has to render her verdict in experiments that either support or falsify the theory's predictions, no matter my intuition.

This is the third book I have written on the topic of experience. *The Quest for Consciousness*, published in 2005, arose out of a class I taught for many years, surveying the vast psychological and neurological literature relevant to subjective experience. I followed this up, in 2012, with *Consciousness: Confessions of a Romantic Reductionist*, which covered scientific advances and discoveries in the intervening years, mixed in with autobiographical overtones. *The Feeling of Life Itself* is devoid of such distractions. All you need to know is that I'm one of seven billion random deals from the deck of human possibilities—I grew up happily, lived in many cities in America, Africa, Europe, and Asia, a physicist turned neurobiologist, a vegetarian cyclist with a soft spot for philosophy and a great love for books, big boisterous dogs, vigorous physical activity, and the outdoors, and a sense of melancholy, living in the twilight of a glorious age.

So let us set off on this voyage of discovery, with consciousness as our lodestar.

Seattle, October 2018

Acknowledgments

Writing books is one of life's great pleasures, rewarding at an intellectual and emotional level over a protracted period, unlike the more fleeting pleasures of the body. Thinking about the book's content, discussing it with others, revising it, and working with editors, artists and the publisher provides a focus for one's mental energies.

I would like to thank everyone who engaged with me over the past three years of writing.

Judith Feldmann took my prose and edited it. Bénédicte Rossi was the artist who turned my cartoons into beautiful drawings. The book's title is an amalgamation of a comment by Elizabeth Koch, who exclaimed after one of my talks that "you study the feeling of life," and the title of Francis Crick's book *Life Itself: Its Origin and Nature*.

Many friends and colleagues read drafts, identified infelicities and inconsistencies and helped me sharpen the underlying concepts. In particular, I would like to acknowledge with gratitude Larissa Albantakis, Melanie Boly, Fatma Deniz, Mike Hawrylycz, Patrick House, David McCormick, Liad Mudrik, and Giulio Tononi, who took the time to carefully read the entire text and emend it. The philosophers Francis Fallon and Matthew Owen helped to clarify some conceptual troubling issues. My daughter, Gabriele Koch, edited key sections. The book is better for all of their efforts.

During the day, I'm the chief scientist and president of the Allen Institute for Brain Science in Seattle, studying the mammalian brain at the cellular level. The science we carry out at our Institute has informed many aspects of this book. I thank the late Paul G. Allen for providing the vision and the means to allow my colleagues and I to tackle hard questions under the motto of "Big Science, Team Science and Open Science." I thank the chief executive officer of the Allen Institute, Allan Jones, for tolerating my scholarly pursuits. I gratefully acknowledge the Tiny Blue Dot Foundation for funding some of the consciousness-related research reported in the book.

Last but not least, I am grateful to my wife, Teresa Ward-Koch who, together with Ruby and Felix, reminds me about what is important in life and for letting me get away with so many late-night and early-morning bouts of solitary writing.

1 What Is Consciousness?

What is common between the delectable taste of a favorite food, the sharp sting of an infected tooth, the fullness after a heavy meal, the slow passage of time while waiting, the willing of a deliberate act, and the mixture of vitality, tinged with anxiety, just before a competitive event?

All are distinct experiences. What cuts across each is that all are subjective states. All are consciously felt. Accounting for the nature of consciousness appears elusive, with many claiming that it cannot be defined at all. Yet defining it is actually straightforward. Here goes:

Consciousness is experience.

That's it. Consciousness is any experience, from the most mundane to the most exalted. Some add *subjective* or *phenomenal* to the definition. For my purposes, these adjectives are redundant. Some distinguish *awareness* from *consciousness*. For reasons I've given elsewhere,[1] I don't find this distinction helpful and so I use these two words interchangeably. I also do not distinguish between *feeling* and *experience*, although in everyday use feeling is usually reserved for strong emotions, such as feeling angry or in love. As I use it, any feeling is an experience. Collectively taken, then, consciousness is lived reality. It is the feeling of life itself. It is the only bit of eternity to which I am entitled. Without experience, I would be a zombie, a nothing to myself.

To be sure, my mind has other aspects, as well. In particular, there is the vast domain of the non- and the unconscious that exists outside the limelight of consciousness. But the challenging part of the mind–body problem is consciousness, not nonconscious processing; it is that I can *see* something, *feel* something, that is mysterious, rather than how my visual system processes the rain of photons impinging onto my retina to identify a face. Any smartphone does the latter, but none can perform the former.

The seventeenth-century French physicist, mathematician, and philosopher René Descartes, in his *Discourse on the Method*, sought ultimate certainty as the foundation of all thought. He reasoned that, if he could assume that everything was open to doubt, including whether the outside world existed, and still know something, then that something would be certain. To this end, Descartes conceived of a "supremely powerful malicious deceiver" who could fool him about the existence of the world, his body, and everything he saw or felt. Yet what was not open to doubt was that he was experiencing *something*. Descartes concluded that because he was conscious, he existed. He expressed this, the most famous deduction in Western thought, in the memorable dictum:

> I think, therefore I am.[2]

More than a thousand years earlier, Saint Augustine of Hippo, one of the foundational Church Fathers, made a strikingly similar argument in his *City of God*, with the tag line, *si fallor sum*, or

> If I am mistaken, I exist.[3]

Less high-brow but closer to contemporary cyberpunk sensibility is Neo, the central character in the *Matrix* movie trilogy. Neo lives in a computer simulation, the Matrix, which looks and feels to him like the everyday "real" world. In reality, Neo's body, together with those of the rest of humanity, is stacked in gigantic warehouses, harvested as an energy source by sentient machines (a modern-day version of Descartes's malicious deceiver). Until Neo takes the red pill offered to him by Morpheus, he lives in complete denial of this reality; yet there is no doubt that Neo has conscious experiences, even though their content is completely delusional.

A different way of putting it is that *phenomenology*—what I experience and how my experiences are structured—is prior to what I can infer about the external world, including scientific laws. Consciousness is prior to physics.

Think of it this way. I see something I've learned to call a face. Face percepts follow certain regularities: they are usually left-right symmetric; they typically consist of something conventionally called a mouth, a nose, two eyes. From closely inspecting the eyes in a face, I can infer whether the face is looking at me, whether it is angry or scared and so on. I implicitly attribute these regularities to objects, called people, existing in a world outside of me; I learn how to interact with them and I infer that I am a person like them. As I grow up, I am so utterly habituated to this inferential process that I take it completely for granted. From these experiences, I build up a picture of the world. This inferential process is amplified and acquires immense power using the intersubjective method of science that reveal hidden aspects of reality, such as electrons and gravity, exploding stars, the genetic code, dinosaurs, and so on. But ultimately, these are all inferences; eminently reasonable ones, but inferences nonetheless. All these things could prove erroneous. But not that I experience. It is the one fact I am absolutely certain of. Everything else is conjecture, including the existence of an external world.

Denying One's Experience

The great strength of this commonsense definition—*consciousness is experience*—is that it is completely obvious. What could be simpler? Consciousness is the way the world appears and feels to me (I will talk about you in the next chapter).

A minority of researchers begs to differ. To reduce the mental discomfort of being unable to explain the central aspect of life, some philosophers, such as the wife-and-husband team of Patricia and Paul Churchland, dismissively refer to the folk belief of the reality of experience as a naive assumption, just like thinking the Earth is flat, that must and shall be overcome. They seek to eliminate the very idea of consciousness from polite discussion among the educated.[4] On this view, in some real sense, no one suffers from cruelty, torture, agony, distress, depression, or anxiety. If correct, such an eliminative stance implies that if people would only realize that they are confused about the true nature of their experiences, that consciousness doesn't really exist, suffering would vanish *tout court* from the world! Utopia achieved (of course, there wouldn't be pleasure and joy either; you can't cook an omelet without breaking some eggs). To put it mildly, I find this extremely unlikely. Such a denial of the authentic nature of experience is a metaphysical counterpart to Cotard's syndrome, a psychiatric condition in which patients deny being alive.

Others, such as Daniel Dennett, argue vociferously that, although consciousness exists, there is nothing intrinsic or special about it. As he expressed it in an interview in the *New York Times*, "The elusive subjective conscious experience—the redness of red, the painfulness of pain—that philosophers call qualia? Sheer illusion."[5] There is nothing real about my excruciating back pain above and beyond my behavioral dispositions, my need to remain absolutely still, flat on the floor, and so on.

These teachings, aided and abetted for self-serving reasons by much of Silicon Valley (which I shall return to in the penultimate chapter), declare the intrinsic nature of consciousness to be the last Grand Illusion that we need to rid ourselves of. I find this absurd—for if consciousness is an illusion shared by all, it remains a subjective experience, an experience no less real than any veridical percept.

Given these eristic arguments, it becomes clear that much of twentieth-century analytic philosophy has gone to the dogs. Indeed, John Searle, the doyen of American philosophers, has these annihilating words for his colleagues:

> The most striking feature of ... mainstream philosophy of mind of the past fifty years ... seems obviously false.[6]

The philosopher Galen Strawson opines:

> If there is any sense in which these philosophers are rejecting the ordinary view of the nature of things like pain ... their view seems to be one of the most amazing manifestations of human irrationality on record. It is much less irrational to postulate the existence of a divine being whom we cannot perceive than to deny the truth of the commonsense view of experience.[7]

I assume that experiences are the only aspect of reality I am directly acquainted with. Their existence poses an obvious challenge to our current, quite limited understanding of the physical nature of reality and they cry out for a rational, empirically testable explanation.

The nineteenth-century physicist Ernst Mach, after whom the speed of sound is named, was an ardent student of phenomenology, the study of the way the world appears to us. I've adapted a famous pencil drawing of his, *Innenperspektive* (fig. 1.1), to make an important point: I don't need a scientific theory, a holy book, the affirmation of any ecclesiastical, political, or philosophical authority, or anybody else to experience something. My conscious experience exists for itself, without the need for anything external, such as an observer. Any theory of consciousness will have to reflect this intrinsic reality.[8]

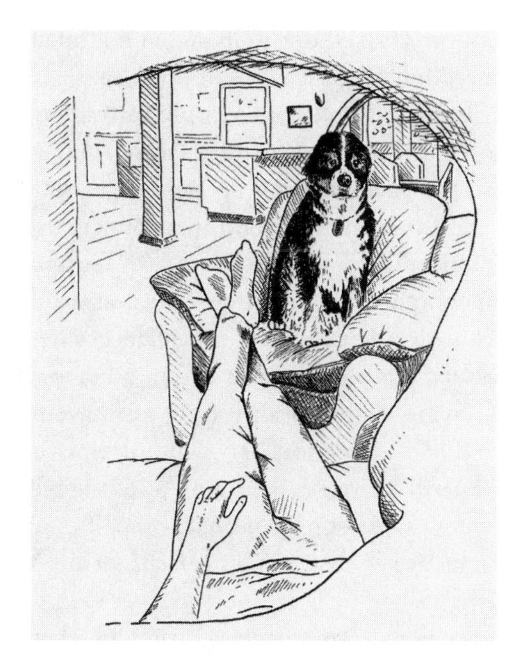

Figure 1.1
Inner perspective: The world as seen through my left eye—including a portion of my eyebrow and nose, and my dog, Ruby, sitting on a lounge chair and looking at me. The extent to which this percept is congruent with reality is ultimately open; perhaps I am hallucinating. But it is a drawing of my conscious visual experience, the only reality I have direct acquaintance with.

The Challenge of Defining Consciousness as Experience

This commonsense definition has one drawback: it only makes sense to other conscious creatures. Explaining experience to a nonconscious superintelligence or to a zombie is meaningless. Whether this will always be so remains to be seen—for what the philosopher Thomas Nagel calls an "objective phenomenology" may be within reach.[9]

In objective terms, *seeing* is closely linked to visuomotor behavior which can be defined as "acting on incoming electromagnetic radiation in a particular part of the spectrum." In this sense, any organism responding to visual input with some action, whether a fly, a dog, or a human, sees. But this description of visuomotor behavior completely leaves out the "seeing" part—the canvas painted with scenes from life, as in figure 1.1. Visuomotor behavior is action—which is all fine in itself, but it is totally different from my subjective perception of the scene in front of me.

These days, it is easy for image-processing software not only to store photos but also to pick out and identify faces. The algorithm extracts information from the pixels making up the image and outputs a label, such as "Mom." Yet this straightforward transformation—image in, label out—is radically different from my experience of seeing my mother. The former is an input–output transformation; the latter is a state of being.

Explaining feelings to a zombie is a much greater challenge than explaining seeing to a person born without sight. For the blind person knows about sounds, touches, loving, hating, and so on; I just have to explain that a visual experience is like an auditory experience except visual percepts are associated with blobs that move in a certain way as the eyes swivel and the head turns and whose surfaces have peculiar properties such as color and texture. The zombie, by contrast, has no percepts of any kind to compare the feeling of seeing to.

I wake up every day to a world suffused with conscious experience. As a rational being, I seek to explain the nature of this luminous feeling, who has it and who doesn't, how it arises out of physics and my body, and whether engineered systems can have it. Just because it is more difficult to define consciousness objectively than to define an electron, a gene, or a black hole, doesn't mean that I have to abandon the quest for a science of consciousness. I just have to work harder at it.

Any Experience Is Structured

Any experience has distinctions within it. That is, any experience is structured, composed of many internal phenomenal distinctions. Consider one particular visual experience (fig. 1.1). Its central focus is my Bernese mountain dog, Ruby, sitting in a chair, onto which my legs are propped. Other objects can be seen in the background. Yet that is not all; there is more, much more. There is left and right, up and down, center and periphery, closer and farther away—an uncountable number of spatial relationships. Even when I open my eyes in the complete dark, I experience a rich notion of geometric space that extends in all directions.

The actual experience, impossible to depict in a drawing, also includes Ruby's peculiar smell and the emotional coloration that shapes my attitude toward her. These distinct sensory and affective aspects are interwoven in a complex experiential cocktail, each with its own time-course, some swift, some more sluggish, some transient, some sustained. This is true of most experiences; each can be dissected into finer distinctions across modalities.[10]

Consider another everyday experience. Squeezed into seat 36F on a bumpy, two-hour flight after having had my morning cappuccino, I feel pressure building up in my bladder. By the time I get to a bathroom in the terminal, the urge to pee becomes almost unbearable[11]—finally, I consciously feel the urine flowing, together with a mildly pleasurable sensation as the pressure is relieved. But beyond that, I can't introspect further. I can't decompose these sensations into more primitive atomic elements. I can't get past the "veil of the Maya," to adopt Hindu parlance. My introspective spade has hit impenetrable bedrock.[12] And I certainly never experience the synapses, neurons and the other stuff inside my skull that constitutes the physical substrate of any experience. That level is completely hidden to me.

Finally, consider a rare class of conscious states: mystical experiences common to many religious traditions, whether Christian, Jewish, Buddhist, or Hindu. These are characterized as having no content: no sounds, no images, no bodily feelings, no memories, no fear, no desire, no ego, no distinction between the experiencer and the experience, the apprehender and the apprehended (nondual).

The late-medieval Dominican monastic, philosopher, and mystic Meister Eckhart encountered the Godhead in a featureless plain, the essence of his soul:

> There is the silent "middle," for no creature ever entered there and no image, nor has the soul there either activity or understanding, therefor she is not aware there of any image, whether of herself or of any other creature.[13]

Using similar language, long-term practitioners of Buddhist meditation describe naked or sheer awareness:

> Unobscured like a cloudless sky, remain in lucid and intangible openness. Unmoving like the ocean free of waves, remain in complete ease, undistracted by thought. Unchanging and brilliant like a flame undisturbed by the wind, remain utterly clean and bright.[14]

I shall return in chapter 10 to content-free or pure consciousness, as this phenomenon constitutes a striking challenge to any computational account of consciousness. Note that even pure experience is, strictly speaking, a subset (though not a proper one) of the whole and is therefore structured.

Beyond the intrinsic and structured nature of any one conscious experience, what else do I know for certain about my experience? What can I positively say that is true for any experience, no matter how mundane or how exotic?

Any Experience Is Informative, Integrated, and Definite

Three additional properties hold for any conscious experience. They cannot be doubted.

First, any experience is highly *informative*, distinct because of the way it is. Each experience is informationally rich, containing a great deal of detail, a composition of specific phenomenal distinctions, bound together in specific ways. Every frame of every movie I ever saw or will see in the future is a distinct experience, each one a wealth of phenomenology of colors, shapes, lines, and textures at locations throughout the field of view. And then there are auditory, olfactory, tactile, sexual, and other bodily experiences—each one distinct in its own way. There cannot be a generic experience. Even the experience of vaguely seeing something in a dense fog, without being clear what I am seeing, is a specific experience.

I recently attended a Blind Café during which I underwent a sort of a reverse birth. I shuffled from a lit vestibule through a long, black, narrow birth-canal into a completely dark chamber—so dark that I was unable to see my wife's hand that she waved in front of me. We groped for chairs, sat down, introduced ourselves to the other guests and started to eat in Stygian darkness—very, very carefully. It was an utterly unique experience designed to introduce sighted folks to the world of the blind. Yet even in this pitch-black room I had a distinct visual experience, specific, and, combined with its echoes and its feels, distinct from waking up in a pitch-dark hotel room.

Second, any experience is *integrated*, irreducible to its independent components. Each experience is unitary, holistic, including all phenomenal distinctions and relations within that experience. I experience the entire drawing, including my body on the couch and the room, not just the legs and, independently, the hand. I don't experience the left side independently of the right side or the dog divorced from the lounge chair on which she is squatting. I experience the whole thing. When somebody tells me about their honeymoon, I have a distinct image of the couple going off on a romantic get-away rather than imagining the sweet substance produced by bees in addition to the large object in the sky.[15]

Third, any experience is *definite* in content and spatiotemporal grain. It is unmistakable. Looking again at the domestic scene in figure 1.1, I perceive my dog and the world in chiaroscuro, in perspective, from the sofa, with my right eye shut. There is a distinct content of consciousness that is "in" while everything else is out, not experienced. The world I see isn't bordered by a line beyond which things are gray or dark, such as behind my head. It simply doesn't exist. The strokes of the brush are painted onto the canvas; everything else is not.

My experience is what it is with a definite content. If it were anything more (seeing while experiencing a pounding headache, say) or anything less (like the drawing but without dog), it would be a different experience.

In summary, every conscious experience has five distinct and undeniable properties: each one exists for itself, is structured, informative, integrated and definite. These are the five essential hallmarks of any and all conscious experiences, from the commonplace to the exalted, from the painful to the orgiastic.

Any Experience Has a Point of View and Occurs in Time

Some researchers argue that experiences may have other properties in addition to these five; for example, that each experience comes with a unique point of view—a first-person account, the subject's perspective. I am looking at the drawing; I am at the center of this world.[16] I suspect that centeredness emerges from the representation of space as it is given to me by my visual, auditory, and tactile senses. Each one of these three associated sensory spaces has one particular location singled out, which is where the eyes, ears, and my body respectively are located. As it is obviously important that what I see, what I hear, and what I feel all refer to a common space (so that, for example, the sounds I hear emanating from moving lips are assigned to the colocalized face), "I" am located at this singular point, the origin of my own space. Furthermore, this center is also the focus of any behaviors, such as moving the eyes with its attendant shift in perspective. Thus, having a perspective, a view from somewhere rather than from nowhere, emerges in a natural way from the structure of sensorimotor contingencies, without having to postulate any additional fundamental property.

A more compelling case could be made that any one experience takes place at a particular moment, the present *now*. Defining this now in an objective manner has defied philosophers, physicists, and psychologists since time immemorial. There is no doubt that lived life has three distinct temporal dominions: the past, the present, and the future, with the experienced present being the intermediate between the past and the future.[17] The past encompasses everything that has already happened. It is immutable, even though how I recall events within my memory palace is susceptible to reinterpretations and to subsequent occurrences that seemingly violate causality. The future is the sum total of everything that hasn't happened yet; it is open-ended and contingent. The bleeding edge of the future forever turns into the specious present that irrevocably recedes into the past as soon as it is experienced.

Yet there are uncommon experiences in which the perception of time ceases. For those taking hallucinogens, for example, the flow of the river of time, the duration of the present now, can slow down and even stop altogether. Time crawls when one's attention is utterly and fully engaged, such as during a dangerous climb up a sheer wall of granite. Movies like *The Matrix* visualize this slowing of perceived time through the well-known bullet-time effect. In other words, the flow of time is not a universal property of all experiences, but only of most.[18]

Thus, what remains is the quintet of essential properties that any and all conscious experiences have:

> Every conscious experience exists for itself, is structured, is the specific way it is, is one, and is definite.[19]

So that's how it is for me. How is it for you? What can I confidently state about the experiences of others? How can their experiences be studied in the laboratory? I cover these questions in the next chapter.

2 Who Is Conscious?

So far, I've talked obsessively about my experiences. I did so because they're the only ones I'm directly acquainted with. This chapter is concerned with your experiences and those of others.

In Roman times, a *privatus* was somebody who had withdrawn from public life (something unimaginable today, in the age of the internet and social media). So it is with any conscious experience: each one is private, inaccessible to anybody else. My perception of seeing yellow is mine and mine alone. Even when you and I both look at the same yellow school bus, you might experience a different hue, and what you experience will almost certainly have different associations for you than what I experience has for me.

This first-person aspect of consciousness is a singular property of the mind, making it more challenging to study than the usual objects that science investigates. For these are defined by properties—mass, motion, electrical charge, molecular structure—accessible to anybody with the appropriate instruments and tools for measurement. Appropriately, they are known as *third-person properties*.

The challenge of the mind–body problem is thus to bridge the divide between the subjective, *first-person perspective* of the experiencing mind and the objective, third-person perspective of science.

Note that other's people experiences are not the only nonobservable entities that science investigates. Most famous is the wave function of quantum mechanics that can't be directly probed. All that is measurable are probabilities derived from the wave function. The multiverse, the vast collection of universes within the cosmos, each one with its distinct physical laws, is another nonobservable entity. Unlike the wave function or consciousness, the multiverse is totally beyond our causal reach, yet remains the object of fevered speculations.[1]

One extreme reaction to the private nature of consciousness is solipsism, the metaphysical doctrine that nothing exists outside my mind. Logically consistent and impossible to disprove, this belief is unproductive, as it doesn't explain interesting facts about the universe I live in. How did my mind arise? Why is it populated with stars, dogs, faces? What laws govern their behavior?

A weaker form of solipsism accepts the reality of the external world but denies the existence of other conscious minds. Everybody but me is a zombie, devoid of feelings, only pretending to love and hate. While logically possible, this idea is intellectual claptrap. For it would suppose that my, and only my, brain gave rise to consciousness. One psychophysical law for my brain, and a different law for the brains of the other seven billion people. The chances of this being true are nil.

To me, solipsism has always seemed an extreme form of egotism, in addition to being sterile and useless. Yes, to please myself I can imagine that I'm the only mind in existence; the moment I die, the world will disappear into the void from which it arose before I first experienced it. But solipsism doesn't explain the world around me. Let us waste no more time on it and get on with the real task.

The Fecundity of Abductive Reasoning

The most rational alternative is to assume that other people, such as you, have conscious experiences. This inference is based on the great similarity of our bodies and brains. It is reinforced if what you tell me about your experiences relates to my experiences in obvious ways.

That you are not a zombie can't be proven on strictly logical grounds. Rather, it is an inference to the best explanation, a form of reasoning that leads to the likeliest explanation of the relevant data. Called *abductive reasoning*, it extrapolates backward to infer the hypothesis that gives the most plausible explanation of all known facts.

Abductive reasoning lies at the core of the scientific enterprise: astronomers of the mid-nineteenth century had noted irregularities in Uranus's orbit. This led the French astronomer Urbain Le Verrier to abduce the existence and location of an unknown planet. Telescopic observations confirmed the existence of Neptune, a triumphal confirmation of Newton's theory of gravity. Darwin and Wallace abduced that evolution by natural selection is the most likely explanation for the distribution of species across ecosystems. Abduction is a form of reasoning that deals with probabilities and likelihoods. The conclusion of a solid abductive argument is a hypothesis that best explains all known facts. We daily abduce the best explanation of a dizzying variety of phenomena—diagnosing the most likely cause of a skin rash, a malfunctioning car, a leaking pipe, a financial or political crisis.

The search for the most likely explanation of all relevant facts is the very opposite mindset of that of conspiracy-minded folks who perceive the malevolent action of their particular Boogieman behind every event (the CIA, Jews, communists). This leads to a contrived, byzantine chain of reasoning, involving the collusion of thousands of individuals, extremely unlikely to have taken place. Sightings of the Virgin Mary in a cheese sandwich, gigantic alien face artifacts on Mars, and the moon-landing conspiracy are all sorry examples of breakdowns of inference to the best explanation.[2]

Sherlock Holmes is a master of abductive reasoning, with the BBC series *Sherlock* visualizing his inferences with vivid graphic overlays. Despite Holmes's claim that he's practicing a "science of deduction," he rarely deduces anything—for that would imply logical necessity. From the two propositions "All men are mortal" and "Socrates is a man," we can deduce by necessity that Socrates will die. In real life, the situation is never that clear. Typically, Holmes abduces the most likely explanation of the facts, as in his celebrated exchange with the police in the short story "Silver Blaze":

> Inspector Gregory: "Is there any point to which you would wish to draw my attention?" Holmes: "To the curious incident of the dog in the night-time." Inspector: "The dog did nothing in the night-time." Holmes: "That was the curious incident."

Holmes abduced that the dog didn't bark because it knew the perpetrator. Abductive reasoning is all the rage in computer science and artificial intelligence, giving software powerful reasoning abilities. An example is IBM's question-and-answering computer system using natural language, Watson, employed in medical diagnostics.[3]

Probing the Conscious Minds of Others

Unlike my own mind, which I am directly acquainted with, I can only abduce the existence of other conscious minds. I can never directly experience them. In particular, I abduce that you and other people have experiences like I do unless I have strong reasons to believe otherwise (say, because they have a brain injury or are severely intoxicated). With this assumption in place,

I can look for systematic connections between consciousness and the physical world.

Psychophysics (literally the "physics of the soul") is the science that seeks to elucidate quantitative relationships between stimuli—a tone, a spoken word, a color field, a picture flashed onto a screen, a heated probe to the skin—and their elicited experiences. A branch of psychology, psychophysics has uncovered reliable, consistent, reproducible, and lawful regularities between objective stimuli and subjective reports.[4]

Although this chapter focuses on seeing, perception is a broad term that includes the traditional five sensory faculties—sight, sound, smell, touch and taste—as well as pain, balance, heartbeat, nausea, and other epigastric sensations.

To quantify phenomenology under laboratory conditions, psychologists don't rely on flowery descriptions. Instead, they ask simple questions. Lots of them. In a typical experiment, volunteers, paid for their time and effort, will stare at a screen while a picture, for instance, a barely visible face or a butterfly superimposed onto a grainy background of checkered light and shadow, is flashed onto a screen. Immediately afterward, they are confronted by the question, "Did you see a face or a butterfly?" (fig. 2.1). Only two answers are permitted, "face" or "butterfly." Not "I'm not sure I saw much" or "Sorry, I don't know." When in doubt, guess.

In practice, subjects don't speak but push buttons on a keyboard, allowing for consistent and rapid action. In this manner, researchers can quickly collect responses from several hundred trials. Pressing a button also tracks the subject's reaction time, which can be mined for further insights.

The experience is reduced to a series of button presses. Averaged over a block of individual trials, such responses are an *objective measure of perception*, as there is a correct answer available to the researchers (since they have access to the computer program that generated the face or butterfly image and so know the right answer). That is, a proverbial third person knows if what the subject reported corresponds to what was on the screen.

While the timing of the button press is easy enough to measure, the swiftness of vision is more difficult to determine. Comparing the timing of EEG signals from your brain when you see the face to the timing when you didn't (as in figure 2.1) indicates that visual experience arises as early as 150 msec or as late as 350 msec after the stimulus enters your eyes.[5]

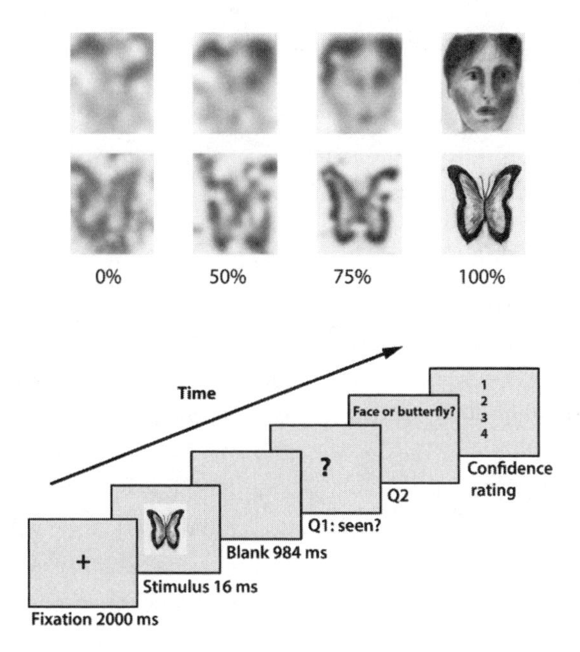

Figure 2.1

Probing experience: Images of faces or butterflies are made easy or more difficult to recognize by overlaying visual noise while you push a button, indicating whether you saw a face or a butterfly. For any one level of such noise, trials when the stimulus is correctly perceived are compared to those when you failed to do so, even though the same picture was present on your retina. (Adapted from Genetti et al., 2011.)

The visibility of the figure is manipulated to make it easier or more difficult to discern. When images are flashed for a mere 1/60th of a second, the perceptual judgment of subjects can vary considerably from trial to trial. Consider the 50 percent face image of figure 2.1 flashed briefly onto the screen. You might respond "face" three times but then on the fourth trial push the butterfly button. As the object emerges from its noisy background to become more visible (75 percent or 100 percent images), you are more and more likely to respond correctly, until you do so on almost every trial. There is a steady progression from not being able to make the distinction, to doing better than chance, to getting it right each time.[6]

Repeating this experiment with many subjects results in similar though not identical response rates as a function of visibility. Responses don't depend much on which pictures are used, butterflies or, say, pictures of animals or houses. This is reassuring and reinforces my prior "We're all conscious" assumption.

Perceptual research of this sort reveals that sensory perception is not a passive reflection or a simple mapping of the outer world onto an inner mental screen. Perception is an active process, "a construction of a description of the world," as the influential theoretician David Marr argued.[7] You are intimately familiar with this world, because it is the one you see, hear, and otherwise experience. You infer this world from the data impinging upon your eyes, ears, and other receptors by sophisticated but unconscious processes. That is, you don't look at the world and say to yourself, "Hmm, that surface reflects light in this way and is occluding another surface, and there is a shadow falling on the first surface cast by another surface far away, with a bright light source on the upper right." No, you look and see a bunch of people under a bright Harvest moon, partially obscuring each other. All of this is inferred based on the available retinal information, as well as your previous visual experiences and those of your ancestors (encoded in your genes).

Perception is a construction of precisely those features that are useful in our struggle to survive a world of eat-or-be-eaten.

How perception happens is hidden from your conscious mind. You simply look and see. Indeed, I still remember when, many decades ago, I tried to explain to my parents—a doctor and a diplomat—why I study vision. They couldn't see the point, as it was so obviously trivial. Similarly, the myriad of software operations underlying even basic tasks on your computer are completely hidden from you behind the simplicity of the user interface.

Visual illusions reveal the sometimes striking dissonance between appearances and reality. Consider the "Lilac chaser," which has its own Wikipedia page (https://en.wikipedia.org/wiki/Lilac_chaser). When you keep your eyes steady on the central fixation cross, you'll see a single greenish disk traveling on a circular trajectory, round and round. Yet the actual stimuli are eleven pink disks on a circle, with the twelfth disk missing in one location; the location of this missing disk, essentially a hole, travels around the circle. What you see is not there, and what is out there on the screen is not what you see!

The Lilac chaser is fool-proof. Even though you know it is an illusion, you can't break it. It is an extreme example of the difference between your perception of the external world and its actual metric properties (size, distance, and so on). Most of the time the conflict between appearance and reality is minor and inconsequential. In that sense, perception is by and large reliable. But at times, the discrepancy can be striking, demonstrating the limits of perception. Even mind-enhancing drugs will not let you escape the cage that is your brain—the world-in-itself, Kant's storied *Ding an sich*, is never directly accessible.

I am an avid rock climber, in search of that peculiar combination of intense fear and exhilaration encountered on the high crag, when time melts into a taut presence. Recently, I was on a narrow rock ledge, high up on a mountain side; it was snowing, with a high wind blowing. I had to cross a chasm on a wooden rope bridge, one side of which was badly frayed. With both feet in full contact with the plank, I slowly and deliberately shuffled across, in control but for the slight shaking of my calf muscles, the "Elvis" or "sewing machine leg" familiar to climbers. Midway between the two walls, over the abyss, I forced myself to look down, at the riverbed far below, before moving on to the relative safety of the narrow ledge on the other side.

Yet in reality, and embarrassingly, I was walking across a wooden stud on the carpeted floor of an office, wearing immersive virtual reality goggles! My visual experience of the cavernous spaces around and below me, my sense of being there, the sound of the wind in my ears—all of it induced a palpable sense of arousal and tension. The abstract knowledge that I was safe did not eliminate the sense of danger I experienced. A visceral demonstration of the limits of perception.

Plumbing the Depths of Consciousness

Psychophysics explores the relation between first-person experiences and third-person objective measures, such as response rates. For some, though, this isn't good enough. They argue that objective measures don't truly capture the subjective nature of an experience. To get closer to the actual phenomenology, psychologists invented *subjective measures*, probing what people know about their experience, a simple form of self-consciousness.

Recall the experiment in which a degraded picture is flashed onto the screen. After you pressed the "face" or "butterfly" button, you are asked to reflect on this button press and indicate how confident you are about your answer. This could take the form of a four-point confidence scale, with a 1 indicating "I'm guessing," a 2 "I may have seen a face," a 3 "I think I saw face," and a 4 "I'm confident I saw a face" (and the same if you answered butterfly). On one trial, you might respond "face, 4" (decoded as "I'm very confident I saw a face"), followed by "butterfly, 2" (decoded as "I may have seen a butterfly"). As the object becomes more clearly visible, both your ability to correctly discriminate the face from the butterfly and your confidence in your judgment increases. The less confident you are about your experience, the worse you do (by objective measures).[8]

Surprisingly, even when you think you're guessing, you can be slightly but significantly better than chance. That is, when confronted with very brief or faint stimuli that don't give rise to a discernible experience, people can still process some of the associated sensory information. Call it a gut feeling. Known as *unconscious priming*, it is highly variable from trial to trial and from experiment to experiment, and has in general only a weak effect on behavior (say, tilting the odds from 50 percent, or chance performance, to 55 percent). Because of the weak and inconsistent nature of unconscious priming, its existence remains controversial.[9]

Such subjective measures have been extended to long questionnaires that ask subjects to rate their experiences along many dimensions on a numerical scale. All aspects of phenomenology can be inventoried in this manner—the strength and timing of visual and other sensory percepts, imagery, memories, thoughts, internal dialogue (the voice inside one's head), self-awareness, cognitive arousal, joy, sexual excitement, feelings of love, anxiety, doubt, oceanic boundlessness, and ego dissolution (the last two for hallucinatory drugs). In this manner, the detailed geography of the mind can be mapped across subjects of diverse gender, ethnicity, and age and probed for differences and similarities.[10]

Before moving on, I must mention one major weakness of existing behavioral techniques that seek to plumb experience. Consider again the experiment of figure 2.1. Depending on the particular image, you see a particular face of a certain age and gender, looking to your right, with a particular facial expression, the eye brows shaped just so, a gray something-or-other on the left cheek, a white mushy pattern on the left, and so on. Each one of these descriptors is a positive distinction. Then there is an abundance of negative distinctions—you are sure you didn't see a cat, a red firetruck, a bunch of letters, nor an uncountable number of other things.

However, none of these positive and negative distinctions get queried by psychologists. The standard psychophysical setup reduces the entire experience to a single distinction, "Did you see a face or a butterfly?" This one-bit answer, akin to that a machine vision classifier, leads to highly reproducible results that are amenable to mathematical analysis. Unfortunately, though, they leave out a universe of distinctions.

Psychologists and philosophers sometimes differentiate between *phenomenal consciousness* and *access consciousness*. The former is what you actually experience while the latter is what you can report, say to the experimentalist.

Some argue that your experience of a scene full of colors, sights, sounds, and fury is illusory, as all that can be accessed are a few simple chunks of data, with the information capacity of consciousness estimated to be between five and nine items, not a lot. The rest is make-believe. Phenomenal consciousness is as impoverished as access consciousness—its content is minute. Yet if all you have to describe your experience is one bit, well, then of course, phenomenology looks terribly impoverished. So the apparent destitution of the content of consciousness is due to inadequate experimental techniques. A facade of poverty hides the fecundity of lush experience. There is more to experience than meets the button![11]

Nonconscious Zombie Agents Run Your Life

While subliminal perception is weak at best, other aspects of the mind are almost always outside the spotlight of consciousness yet affect you in a robust manner. This is the realm of the nonconscious (I avoid the term "unconscious" here, given its strong Freudian connotations).[12]

Whether you're driving, watching media, or talking to friends, you constantly shift your eyes in a series of fast and jerky eye movements called *saccades*. Although this happens three to four times every second of your waking life, you are almost never aware of these incessant movements.

Consider what would happen if you were to move your smartphone camera at the same cadence while snapping a picture. That's right—the picture would come out all blurry. How can your visual world look so clear, without any motion-made smears, when your image sensors, your eyes, are constantly moving? The answer is that your nonconscious mind edits out these blurred segments, a trick known as saccadic suppression. Indeed, you can't ever catch your own eyes in the act of moving. Look into a mirror while rapidly scanning your eyes back and forth—you will see your eyes here, and then there, but never in between. A friend looking on can perfectly well see your eyes move; you just can't see your own. Your brain suppresses these short segments of what would look blurry and replaces them by splicing in a stationary scene, like in a movie studio. The same is true of the eye *blinks* you make every couple of seconds (this editing-out doesn't occur for voluntary *winks*). All this furious editing goes unnoticed; you see a steady world as you look around.

Given that you make more than 100,000 daily saccades, each one lasting between 20 and 100 milliseconds, saccadic and blink suppression adds up to more than an hour a day during which you are effectively blind! Yet until scientists started studying eye movements, no one was aware of this remarkable fact.

Eye movements are but one instance of a sophisticated set of processes, implemented by specialized brain circuits, that make up a lived life. Neurological and psychological sleuthing has uncovered a menagerie of such specialized processes. Hitched to the eyes, ears, the equilibrium organ, and other sensors, these servomechanisms control our eyes, neck, trunk, arms, hands, fingers, legs, and feet. They are responsible for everyday actions—shaving, washing, tying shoelaces, biking to work, typing on a computer keyboard, texting on a phone, playing soccer, and on and on. Francis Crick and I called these specialized sensory-cognitive-motor routines *zombie agents*.[13] They manage the fluid and rapid interplay of muscles and nerves that is at the heart of all skills. They resemble *reflexes*—blinking, coughing, jerking your hand away from a hot stove, or being startled by a sudden loud noise. Classical reflexes are automatic, fast, and depend on circuits in the spinal cord or in the brainstem. Zombie behaviors can be thought of as more flexible and adaptive reflexes that involve the forebrain.

Saccadic eye movements are controlled by such a zombie agent while bypassing consciousness. You can become conscious of the routine action of a zombie agent, but only after the fact. While I was trail running in the mountains in Southern California, something made me look down. My right leg instantly lengthened its stride, for my brain had detected a rattlesnake sunning itself on the stony path where I was about to put my foot. Before I had consciously seen the reptile, before I had experienced the attendant adrenaline rush, and before the snake gave its ominous warning rattle—I had avoided stepping on it and sped past. If I had depended on the conscious feeling of fear to control my legs, I would have trod on the snake. Experiments confirm that motor action can, indeed, be faster than thought, with the onset of corrective motor action preceding conscious perception by about a quarter of a second. Likewise, consider a world-class sprinter running one hundred meters in ten seconds. By the time the runner consciously hears the pistol, he is already several strides out of the starting block.

Learning a new sport—playing tennis, sailing, sculling, or mountaineering—takes a great deal of both physical and mental discipline. In climbing, the novice learns where to place her hands, feet, and body to smear, stem, lie back, lock off the wrist or the fingers in a crack. The climber pays attention to the flakes and grooves that turn a vertical granite cliff into a climbable wall with holds and learns to ignore the void beneath her. A sequence of distinct sensory-cognitive-motor actions is stitched into a smoothly executing motor program. After hundreds of hours of intense and dedicated training these labors result in thoughtless, flawless flow, a divine experience. Constant repetition recruits specialized brain circuits, colloquially referred to as *muscle memory*, that render the skill effortless, the motion of the body fluid, without wasted effort. The expert climber never gives a thought to the minutiae of her actions requiring a marvelous merging of muscle and nerve.

Indeed, much of what goes on in unexamined life is not accessible to consciousness or bypasses it altogether. A widespread phenomenon is *mind blanking*: the mind is seemingly nowhere while the body carries on with the routine aspects of everyday living.[14] Virginia Woolf, an astute observer of the inner self, had this to say:

> Often when I have been writing one of my so-called novels I have been baffled by this same problem; that is, how to describe what I call in my private shorthand "non-being." Every day includes much more non-being than being.... Although it was a good day the goodness was embedded in a kind of nondescript cotton wool. This is always so. A great part of every day is not lived consciously. One walks, eats, sees things, deals with what has to be done; the broken vacuum cleaner.... When it is a bad day the proportion of non-being is much larger.[15]

Just as you can never catch the light inside your refrigerator turned off, because every time you open the door of the fridge the light is on, you can't experience not experiencing. By randomly pinging subjects on their smartphones to ask what they were aware of at that precise point in time, psychologists discovered that the blank mind, defined by no experience at all (quite the opposite of pure experience mentioned in the previous chapter), is common over the course of a day at the office, doing chores at home or while at the gym, driving, or watching TV. Mindfulness, "being in the moment," counteracts the blank mind.

Somewhere in the brain, the body is monitored; love, joy, and fear are born; thoughts arise, are mulled over, and discarded; plans are made; memories are laid down. Yet the conscious self may be turned off or is oblivious to this furious activity. You are a stranger to your mind.

That most of the operations of the mind are inaccessible to consciousness should not be surprising. After all, you don't feel your liver metabolizing the alcohol in the pinot noir from last night, you don't experience the trillions of bacteria happily colonizing your intestine, and you are deaf to your immune system fighting off some bug.

The nonconscious is a bound of the mind that lay undiscovered until philosophers and psychologists—in particular, Friedrich Nietzsche, Sigmund Freud, and Pierre Janet—inferred its existence in the late 1800s. Its remoteness reflects our deeply felt intuition that the conscious mind is all there is. It also explains why so much of philosophy of mind has been barren. You can't introspect your way to the unconscious layers of your mind. Yogi Berra might have quipped "You don't know what you don't experience."

The existence of the nonconscious throws the question of the physical basis of consciousness into stark relief. What is the difference between unconscious and conscious actions of the mind?

On the Limits of Behavioral Methods

You might think that scientists wouldn't touch any measure characterized as subjective with a ten feet pole. Yet subjective doesn't mean arbitrary. Subjective measures follow well-established regularities that can be checked. As a rule, as stimulus duration or the strength of the central object relative to its background decreases (fig. 2.1), both objective response rate and subjective confidence decrease and reaction time lengthens—the less certain you are about what you experienced, the slower you respond. In other words, the first-person perspective can be validated by third-person measures.

It is good scientific practice to assume that volunteers may not always faithfully carry out the instructions of the experimentalist—either because they can't follow instructions (as is the case with babies and infants, who require special techniques), misunderstand them, or don't want to follow them (because subjects are bored and press buttons at random or want to cheat). It is critical to design appropriate controls; adding catch trials where the answer is known, repeating experiments to check for consistency, and cross-validating with other data to keep such inappropriate responses to a minimum.

There are subjects, however, who are as isolated as any early twentieth-century polar explorer wintering in the arctic—patients with severe disorders of consciousness following traumatic brain injury, encephalitis, meningitis, stroke, drug or alcohol intoxication, or cardiac arrest. Disabled and bedridden, they are unable to talk about or otherwise signal their mental state. Unlike comatose patients who have few reflexes and lie immobile, in a profound state of unconsciousness, vegetative-state patients cycle through periods of eyes opening and closing, resembling sleep (without necessarily having sleep-associated brain-wave activity).[16] They may move their limbs reflexively, grimace, turn their head, groan, spasmodically move their hands. To the naive bedside observer, these movements and sounds suggest that the patient is awake, desperately trying to communicate with his or her loved ones.

Consider Terri Schiavo, the woman in Florida who lingered for fifteen years in a vegetative state until her medically induced death in 2005. Given the public fight between her husband, who advocated discontinuing her life support, and her devout parents, who believed that their daughter had some measure of awareness, the case caused a huge uproar. It was litigated up and down the judicial chain and eventually landed on the desk of then-president George W. Bush. Her husband ultimately prevailed in his wish to have his wife taken off life support.[17]

Properly diagnosing vegetative-state patients is challenging. Who can say with certainty whether these patients experience pain and distress, living in the gray zone between fleeting consciousness and nothingness? Fortunately, neurotechnology is coming to the rescue of such patients, as I shall detail in chapter 9.

Thus, the absence of reproducible, willful behavior is not always a sure sign of unconsciousness. Conversely, the presence of some behaviors is likewise not always a sure sign of consciousness. A variety of reflex-like behaviors—eye movements, posture adjustments, or mumbling in one's sleep—bypass consciousness. Sleepwalkers are capable of complex, stereotyped behaviors—moving about, dressing and undressing, and so on, without subsequent recall or other evidence of awareness.[18]

So yes—behavioral methods of inferring experience in others do have limitations; but even these limitations can be studied objectively. And as science's understanding of consciousness grows, the frontier between the known and the unknown is constantly being pushed back.

So far, I have only spoken about people and their experiences. What about animals? Do they too see, hear, smell, love, fear, and grieve?

3 Animal Consciousness

The contrast could not have been starker—here was one of the world's most revered figures, His Holiness the Fourteenth Dalai Lama, expressing his belief that all life is sentient, while I, as a card-carrying neuroscientist, presented the contemporary Western consensus that some animals might, perhaps, possibly, share the precious gift of sentience, of conscious experience, with humans.

The setting was a symposium between Buddhist monk-scholars and Western scientists in a Tibetan monastery in Southern India, fostering a dialogue in physics, biology, and brain science.[1]

Buddhism has philosophical traditions reaching back to the fifth century BCE. It defines life as possessing heat (i.e., a metabolism) and sentience, that is, the ability to sense, to experience, and to act. According to its teachings, consciousness is accorded to all animals, large and small—human adults and fetuses, monkeys, dogs, fish, and even lowly cockroaches and mosquitoes. All of them can suffer; all their lives are precious.

Compare this all-encompassing attitude of reverence to the historic view in the West. Abrahamic religions preach human exceptionalism—although animals have sensibilities, drives, and motivations and can act intelligently, they do not have an immortal soul that marks them as special, as able to be resurrected beyond history, in the Eschaton. On my travels and public talks, I still encounter plenty of scientists and others who, explicitly or implicitly, hold to human exclusivity. Cultural mores change slowly, and early childhood religious imprinting is powerful.

I grew up in a devout Roman Catholic family with Purzel, a fearless dachshund. Purzel could be affectionate, curious, playful, aggressive, ashamed, or anxious. Yet my church taught that dogs do not have souls. Only humans do. Even as a child, I felt intuitively that this was wrong; either we all have souls, whatever that means, or none of us do.

René Descartes famously argued that a dog howling pitifully when hit by a carriage does not feel pain. The dog is simply a broken machine, devoid of the *res cogitans* or cognitive substance that is the hallmark of people. For those who argue that Descartes didn't truly believe that dogs and other animals had no feelings, I present the fact that he, like other natural philosophers of his age, performed vivisection on rabbits and dogs.[2] That's live coronary surgery without anything to dull the agonizing pain. As much as I admire Descartes as a revolutionary thinker, I find this difficult to stomach.

Modernity abandoned the belief in a Cartesian soul, but the dominant cultural narrative remains—humans are special; they are above and beyond all other creatures. All humans enjoy universal rights, yet no animal does. No animal possesses the fundamental right to its life, to bodily liberty and integrity. I will return to this grim state of affairs in the book's coda.

Yet the same abductive inference used to infer experience in other people can also be applied to nonhuman animals. In this chapter, I will specifically address the question of experience in mammals, our kind.[3] The last chapter will consider the extent to which consciousness can be found in evolved creatures other than mammals.

Genetic, Physiological, and Behavioral Continuity

I am confident in abducing experiences in fellow mammals for three reasons.

First, all mammals are closely related evolutionarily speaking. Placental mammals trace their common descent to small, furry, nocturnal creatures that scurried the forest in search of insects. After an asteroid killed off most of the remaining dinosaurs about 65 million years ago, mammals diversified and occupied all those ecological niches that were swept clean by this planet-wide catastrophe.

Modern humans are genetically most closely linked to chimpanzees. The genomes of these two species, the instruction manual for how to assemble these creatures, differ by only one out of every hundred words.[4] We're not that different from mice either, with almost all mouse genes having a counterpart in the human genome. Thus, when I write "humans and animals," I simply respect dominant linguistic, cultural, and legal customs in differentiating between the two natural kinds, not because I believe in the non-animal nature of humans.

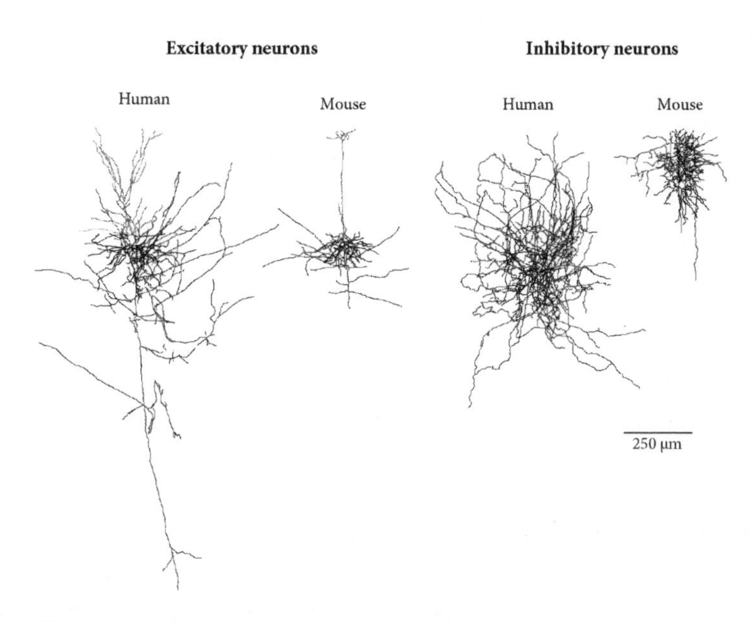

Figure 3.1
Mouse and human neurons: Two human and two mouse neocortical neurons from the Allen Institute of Brain Science. Their morphologies are similar, except that human cells are taller. (Data provided by Staci Sorensen at the Allen Institute.)

Second, the architecture of the nervous system is remarkably conserved across all mammals. Most of the close to nine hundred distinct annotated macroscopic structures found in the human brain are present in the mouse brain, the animal of choice for experimentalists, even though it is a thousand times smaller.[5]

It is not easy, even for a neuroanatomist armed with a microscope, to distinguish human nerve cells from their murine counterpart, once the scale bar has been removed (fig. 3.1).[6] That's not to say human neurons are the same as mouse neurons—they are not; the former are more complex, have more dendritic spines, and look to be more diverse than the latter. The same story holds at the genomic, synaptic, cellular, connectional, and architectural levels—we see a myriad of quantitative but no qualitative differences between the brains of mice, dogs, monkeys, and people. The receptors and pathways that mediate pain are analogous across species.

The human brain is big, but other creatures, such as elephants, dolphins, and whales, have bigger ones. Embarrassingly, some not only have a larger neocortex but also one with twice as many cortical neurons as humans.[7]

Third, the behavior of mammals is kindred to that of people. Take Ruby—she loves to lick the remaining cream off the whisk I use to whip heavy cream by hand—no matter where she is in the house or garden she comes running in as soon as she hears the sounds of the metal wire loops striking the glass. Her behavior tells me that she enjoys the sweet and fatty whipping cream as much as I do; I infer that she has a pleasurable experience. Or when she yelps, whines, gnaws at her paw, limps, and then comes to me, seeking aid: I infer that she's in pain because under similar conditions I act similarly (sans gnawing). Physiologic measures confirm this inference: dogs, just like people, have an elevated heart rate and blood pressure and release stress hormones into their bloodstream when in pain. Dogs not only experience pain from physical injuries but can also suffer, for example if they are beaten or otherwise abused or when an older pet is separated from its litter mate or its human companion. This is not to argue that dog-pain is identical to people-pain; it is not. But all the evidence is compatible with the supposition that dogs, and other mammals, not only react to noxious stimuli but also experience the awfulness of pain and suffering.

As I write this chapter, the world is witnessing a killer whale carrying her baby calf, born dead, for more than two weeks and a thousand miles across the waters of the Pacific Northwest. As the corpse of the baby orca keeps on falling off and sinking, the mother has to expend considerable energy to dive after it and retrieve it, an astonishing display of maternal grief.[8]

Monkeys, dogs, cats, horses, donkeys, rats, mice, and other mammals can all be taught to respond to forced-choice experiments of the sort outlined earlier—modified from those used by people to accommodate paws and snouts, and using food or social rewards in lieu of money. Their responses are remarkably similar to the way people behave, once differences in their sensory organs are accounted for.[9]

Experience without Voice

The most obvious trait that distinguishes humans from other animals is language. Everyday speech represents and communicates abstract symbols and concepts. It is the bedrock of mathematics, science, civilization, and all of our cultural accomplishments.

Many classical scholars assign to language the role of kingmaker when it comes to consciousness. That is, language use is thought to either directly enable consciousness or to be one of the signature behaviors associated with consciousness. This draws a bright line between animals and people. On the far shore of this Rubicon live all creatures, small and large—bees, squids, dogs, and apes; while they have many of the same behavioral and neuronal manifestations of seeing, hearing, smelling, and experiencing pain and pleasure that people have, they have no feelings. They are mere biological machines, devoid of any inner light. On the near shore of this Rubicon lives a sole extant species, *Homo sapiens*.[10] Somehow, the same sort of biological stuff that makes up the brains of creatures across the river is superadded with sentience (Descartes's *res cogitans* or the Christian soul) on this side of the Rubicon.

One of the few remaining contemporary psychologists who denies the evolutionary continuity of consciousness is Euan Macphail. He avers that language and a sense of self are necessary for consciousness. According to him, neither animals nor young children experience anything, as they are unable to speak and have no sense of self—a remarkable conclusion that must endear him to parents and pediatric anesthesiologists everywhere.[11]

What does the evidence suggest? What happens if somebody loses their ability to speak? How does this affect their thinking, sense of self, and their conscious experience of the world? *Aphasia* is the name given to language disorders caused by limited brain damage, usually but not always to the left cortical hemisphere. There are different forms of aphasia—depending on the location of the damage, it can affect the comprehension of speech or of written text, the ability to properly name objects, the production of speech, its grammar, the severity of the deficit, and so on.[12]

The neuroanatomist Jill Bolte Taylor rocketed to fame on the strength of her TED Talk and a subsequent bestselling book about her experience while suffering a stroke.[13] At age thirty-seven, she suffered a massive bleeding in her left hemisphere. For the next several hours, she became effectively mute. She also lost her inner speech, the unvoiced monologue that accompanies us everywhere, and her right hand became paralyzed. Taylor realized that her verbal utterances did not make any sense and that she couldn't understand the gibberish of others. She vividly recalls how she perceived the world in images while experiencing the direct effect of her stroke, wondering how to communicate with people. Hardly the actions of an unconscious zombie.

Two objections to Taylor's compelling personal story is that her narrative can't be directly verified—she suffered the stroke at home, alone—and that she reconstructed these events months and years after the actual episode. Consider then the singular case of a forty-seven-year-old man with an arteriovenous malformation in his brain that triggered minor sensory seizures. As part of his medical workup, regions in his left cerebral hemisphere were anesthetized by a local injection. This lead to a dense aphasia, lasting for about ten minutes, during which he was unable to name animals, answer simple yes/no questions, or describe pictures. When asked to write down immediately afterward what he recalled, it became apparent that he was cognizant of what was happening:

> In general my mind seemed to work except that words could not be found or had turned into other words. I also perceived throughout this procedure what a terrible disorder that would be if it were not reversible due to local anesthetics. There was never a doubt that I would be able to recall what was said or done, the problem was that I often could not do it.[14]

He correctly recalled that he saw a picture of a tennis racket, recognized it for what it was, making the gesture of holding a racket with his hand and explaining that he had just bought a racket. Yet in actuality all he said was "perkbull." What is clear is that the patient continued to experience the world during his brief aphasia. Consciousness did not fade with the degradation of his or Jill Taylor's linguistic skills.

There is ample evidence from split-brain patients that consciousness can be preserved in the nonspeaking cortical hemisphere, usually the right one. These are patients whose corpus callosum (fig. 10.1) has been surgically cut to prevent aberrant electrical activity from spreading from one to the other hemisphere. Almost half a century of research demonstrates that these patients have two conscious minds. Each cortical hemisphere has its own mind, each with its own peculiarities. The left cortex supports normal linguistic processing and speech; the right hemisphere is nearly mute but can read whole words and, in some cases at least, can understand syntax and produce simple speech and song.[15]

It could be countered that language is necessary for the proper development of consciousness but that once this has taken place, language is no longer needed to experience. This hypothesis is difficult to address comprehensively, as it would require raising a child under severe social deprivation.

There are documented cases of feral children who either grew up in near total social isolation or who lived with groups of nonhuman primates, wolfs or dogs. While such extreme abuse and neglect leads to severe linguistic deficits, it does not deprive these wild children of experiencing the world, usually in a tragic and, to them, incomprehensible, manner.[16]

Finally, to restate the obvious—language contributes massively to the way we experience the world, in particular to our sense of the self as our narrative center in the past and present. But our basic experience of the world does not depend on it.

Besides true language, there are, of course, other cognitive differences between people and the rest of mammals. Humans can organize into vast and flexible alliances to pursue common religious, political, military, economic, and scientific projects. We can be deliberately cruel. Shakespeare's Richard III spits out:

> No beast so fierce but knows some touch of pity. But I know none, and therefore am no beast.

We can also introspect, second-guessing our actions and motivations. As we grow up, we acquire a sense of mortality, a knowledge that our life has a finite horizon, the worm at the core of human existence. Death has no such dominion over animals.[17]

The belief that only humans experience anything is preposterous, a remnant of an atavistic desire to be the one species of singular importance to the universe at large. Far more reasonable and compatible with all known facts is the assumption that we share the experience of life with all mammals. How far consciousness reaches down the tree of life will be taken up in the closing chapter.

Before I come to the neuroscience and biology of consciousness, I have one remaining challenge. Many mental operations are thought to be closely related to experience. This applies in particular to thought, intelligence, and attention. Let me now discuss why these cognitive routines can and should be distinguished from consciousness. Experience is different from thinking, being smart, or attending.

4 Consciousness and the Rest

The history of any scientific concept, such as energy, memory, or genetics, is one of increasing differentiation and sophistication until its essence can be explained in a quantitative and mechanistic manner. Over the last few decades, this process of clarification has occurred for experience, a concept that is often conflated with other functions that the mind routinely performs, such as speaking, attending, remembering, or planning.

I will discuss both old and new observations that indicate that experience, though often associated with thought, intelligence, and attention, is distinct from these processes. That is, although consciousness is often entangled with all three cognitive operations, it can be dissociated from them. These findings clear the deck for a concerted attack on the core problem: identifying the neural causes of consciousness and explaining why the brain, but not other organs, gives rise to experience.

Consciousness and the Information-Processing Pyramid

Historically, consciousness has been associated with the most rarefied aspects of the mind. Its information-processing hierarchy is often compared to a pyramid. At the bottom are the massive parallel peripheral processes that register the incoming streams of photons in the retina, changes in air pressure in the cochlear, molecules binding to chemical receptors in the olfactory epithelium, and so on, and convert them into low-level visual, auditory, olfactory, and other sensory events. These are further processed in intermediate stages of the brain until they are turned into abstract symbols in the upper echelons of the mind—you see your friend and hear her ask a question. At the apex of the information-processing hierarchy are powerful cognitive abilities—speaking, symbolic thinking, reasoning, planning, introspecting—the higher "psychical faculty" that only humans and, to a lesser extent, the great apes possess. These capabilities have a limited bandwidth, in the sense that only little data can be processed at a time at this top level.[1] On this view, only a few elite species attain the rank of being conscious and only for the most sophisticated tasks.[2]

However, over the past century the scientific perspective on consciousness has undergone a curious inversion: consciousness has been evicted from the tip of this processing pyramid and has migrated downward. There is nothing refined, reflective, or abstract about an itchy nose, a throbbing headache, the smell of garlic, or a view of the blue sky. A vast multitude of experiences have this elementary character, above the level of raw data streams from exteroceptive (vision, sound, olfactory) and interoceptive (pain, temperature, gut, and other bodily) sensors at the base of the information-processing pyramid but below the highly elaborated, symbolic, and sparse stages at its top. If this is the true state of affairs, then it is overwhelmingly likely that not only humans but many and perhaps most animals, small and large, experience the world.

Indeed, it turns out that our most refined cognitive abilities, such as thinking or being creative, are not even directly accessible to experience. Consider an everyday situation. I'm getting ready to travel and, unbidden, the thought "I need to make a reservation for the three o'clock ferry to Lopez Island" pops into my head. I'm aware of a ghostly juxtaposition of images of the face of a clock with the pointer at three, the ferry, the ocean and the islands, and a speechless inner voice with an injunction to find time to make the online reservation. And this inner speech comes with syntactic and phonological structure.

Life is filled with an astonishing quantity of such linguistic imagery—an inner voice that speculates, plans, admonishes, and comments on events. Only intense physical activity, acute danger, meditation, or deep sleep quiets this constant companion (one reason why rock climbing, biking through dense traffic, reconnoitering enemy terrain, and other physical and cognitively demanding exertions where failure comes with immediate and significant consequences can induce a profound sense of peace—a heard silence inside the mind). The cognitive linguist Ray Jackendoff[3] argues at length that thoughts are represented and manipulated at a semantic level of meaning inaccessible to experience (as did Sigmund Freud).[4] Think of the tip-of-the-tongue phenomenon—you are on the verge of saying a name or concept but can't find the right word, even though you sometimes even have the image in mind. The meaning is there, implicitly, but not the sounds, the phonological structure.

Quite a striking picture emerges from this insight—you are only conscious of the reflections of the outer world in terms of visual, auditory, and other spaces. Likewise, you are only conscious of the reflections of your inner world of mentation onto similar seen or heard spaces. There is a pleasing symmetry to this view—the experience of both outer and inner worlds is primarily sensory-spatial in character (visual, auditory, bodily, and so on) rather than abstract or symbolic (fig. 4.1).

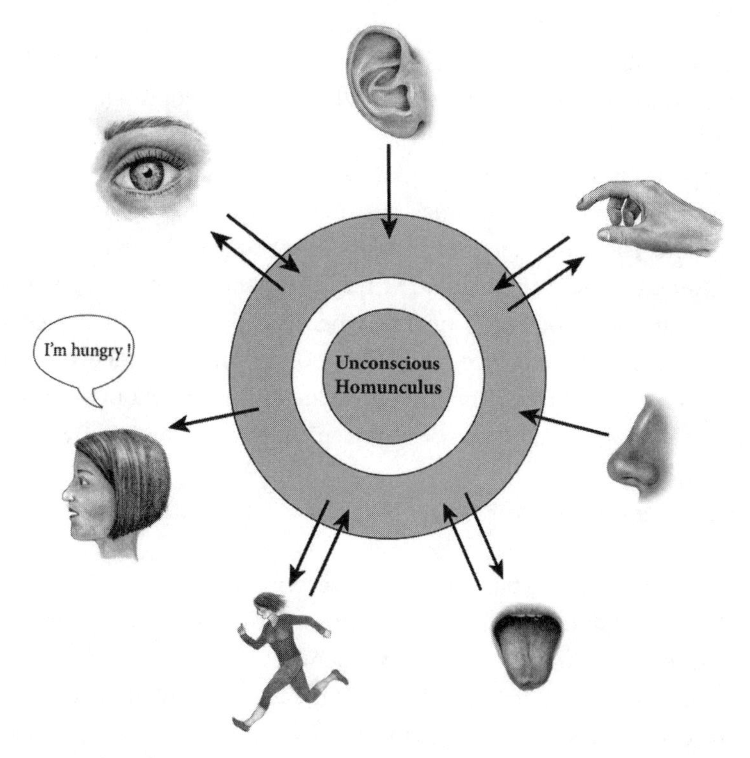

Figure 4.1

The unconscious homunculus: You are conscious neither of raw sensory data, whether originating outside or from within your body, nor of the highest processing stages of the mind, what Francis Crick and I call the unconscious homunculus, the inner source of creativity, thought, and intelligence. The vast majority of your experiences are sensory-spatial in character (white annulus). The arrows denote sensory and motor pathways that connect the brain to the world.

This hypothesis accounts for a compelling illusion, the persistent feeling that there is a little person, a homunculus, inside your head who looks out at the world, thinks, plans, and initiates the action of the sovereign "I." Frequently ridiculed, the idea of a homunculus is, nevertheless, profoundly appealing because it resonates with the everyday experience of who you are.[5] This *unconscious homunculus* (fig. 4.1) is responsible for creativity, intelligence, and planning, much of which is unconscious.

Consider scientific and artistic creativity, the ability to fashion something novel out of existing styles, ideas, and concepts. Jacques Hadamard queried famous scientists and fellow mathematicians about the origins of their innovative ideas. They reported that a long time of intense engagement with their particular problem, an incubation period, followed by a good night's sleep or a few day's diversions, preceded the crucial insight that just "popped into their head." The cognitive inaccessibility of insights has been confirmed by more recent studies.[6]

Creativity and insight are two key aspects of intelligence. If these are inaccessible to conscious introspection, then the relationship between intelligence and consciousness is not straightforward. Maybe these are really two different aspects of the mind? Isn't intelligence ultimately about acting smart in the world and surviving, whereas experience is about feeling? Under that view, intelligence is all about doing, while experience is about being. I shall return to this important theme in the context of animal and machine intelligence and consciousness in chapters 11 and 13 (fig. 13.4).

Consciousness and Attention

I hope you've paid attention to what I have written so far; if you have, you'll visualize this homunculus, living deep within the recess of your mind. If you haven't, these words will have entered your eyes but will have sunk without leaving a trace—you didn't attend to them because your mind was elsewhere.

A teacher reminds her students to pay attention or a psychologist asks his subjects to attend to certain part of an image. What is this "attention" that needs to be called upon for the mind to apprehend certain events, objects, or concepts? Is attention the critical antechamber to the sanctum of consciousness? Can you experience an object or event without attending to it?

There are many forms of attention such as saliency-based, automatic attention, spatial and temporal attention, and feature- and object-based attention. Common to all is that they provide access to processing resources that are in short supply. Because of the limited capacity of any nervous system, no matter how large, it can't process all of the incoming streams of data in real time. Instead, the mind concentrates its computational resources on any one particular task, such as part of a scene unfolding in front of your eyes, and then switches to focus on another task, such as a simultaneously ongoing conversation. Selective attention is evolution's answer to information overload. Its actions and properties have been investigated in considerable detail in the mammalian visual system for more than a century.

Many striking effects demonstrate that if you don't attend to an event you can miss it, even if you are looking directly at it. Consider the "gorilla in our midst" illusion, in which naive subjects track a ball in an ongoing basketball game while a man dressed in a gorilla suit slowly strides across the court—many people completely fail to see the gorilla. Known as *inattentional blindness*, such remarkable failures of vision are much more likely to occur when speaking or texting on a cell phone, which is why driving while using a phone can cause so much mayhem and is illegal in many places.[7]

Thus, visual experiences can depend critically on selective attention. When you pay attention to an object, you usually become conscious of its various attributes; when you shift your attention away, the object fades from your consciousness. This has prompted many to posit that these two processes are inextricably interwoven, if not identical. Going back to the nineteenth century, others, however, have argued that attention and consciousness are distinct phenomena, with separate functions and neuronal mechanisms.

Experimental psychologists investigate attention without consciousness using perceptually invisible stimuli. For example, images of male and female nudes attract spatially selective visual processing (a.k.a. attention) even though they are rendered completely invisible via a cinematographic technique psychologists call masking. But they are still processed, depending on their gender and the sexual orientation of the subject. Such experiments have been repeated in many different contexts, demonstrating that you can attend to objects or events without you becoming conscious of them.[8] Mind blanking while working, eating, or driving is another, albeit less well-studied, instance of attention without consciousness.

While there is a broad consensus that attending to something does not guarantee that it is consciously experienced, the existence of the opposite dissociation, consciousness without attention, is more controversial.

However, when I attend to a particular location or object, intently scrutinizing it, the rest of the world is not reduced to a tunnel, with everything outside the focus of attention fading away: I am always aware of some aspects of the world surrounding me. I am aware that I am looking at text, or driving along a freeway with an overpass approaching.

"Gist" refers to a compact, high-level summary of a scene—a traffic jam on a freeway, crowds at a sports arena, a person with a gun, and so on. Computing gist does not require attentional processing: when a large photograph is briefly and unexpectedly flashed onto a screen while you're focusing on some itsy-bitsy detail, you still apprehend the gist of the photo. A glimpse lasting only one-twentieth of a second is all it takes. And in that brief time, attentional selection does not play much of a role.[9]

Certain complex sensorimotor tasks can be carried out simultaneously, such as driving long stretches on a freeway while listening to an engaging podcast or radio show. Given attentional limitations and the time and cognitive effort required to switch from monitoring the road to following the narrative and back, attention is allocated to only one of these tasks. Yet even when following the storyline, the visual scene in front of me does not fade. I continue to see.[10] Thus, I favor the supposition that selective attention is neither necessary nor sufficient to experience something. Demonstrating this may require subtle manipulations of neural circuits that mediate top-down forms of attention in experimental animals and, ultimately, in people. Stay tuned to the relevant literature.

In summary, dissociating consciousness from language (discussed in the last chapter), thought, intelligence, and attention does not necessarily imply that consciousness isn't heavily entangled with these processes. As I type this sentence, my conscious gaze shifts from the dog at my feet, to a book on my kitchen counter, to foggy Lake Washington outside the window. As I attend to each item, one after the other, I become aware of them, I can take account of them when planning the rest of my day, and so on. Yet these operations—language, attention, memory, planning—can be distinguished from raw experience. Therefore, they will have distinct, but possibly overlapping, physical mechanisms that support them. Of course, in many cases, computers can already speak, attend, remember, and plan. It is experience that remains inexplicable.

But enough of the experiencing mind. Let me turn to the principal organ that supports the mind: the brain.

5 Consciousness and the Brain

Today, we know that the ghost that we give up at death is intimately tied up with three pounds of the tofu-like organ cradled inside its protective bone-case. But this was not always the case.

I here trace the dawning of the neuro-centric age, the critical distinction between states of consciousness and conscious states, and the logic underlying the search for the neuronal footprints of consciousness.[1]

From the Heart to the Brain

For much of recorded history, the heart was considered the seat of reason, emotion, valor, and mind. Indeed, the first step in mummification in ancient Egypt was to scoop out the brain via the nostrils and discard it while the heart, the liver and other internal organs were carefully extracted and preserved so that the Pharaoh had access to everything he needed in his afterlife, everything except for his brain!

Several millennia later, the Greeks didn't do much better.[2] Plato was, as ever, resolute against any empirical investigations of such matters, preferring the Socratic dialogue. Unfortunately, because so much of the mind operates outside the limelight of consciousness, deducing its properties by cogitation and lucubration has proven relatively barren.

Aristotle, one of the greatest of all biologists, taxonomists, embryologists, and the first evolutionist, wrote:

> And of course, the brain is not responsible for any of the sensations at all. The correct view [is] that the seat and source of sensation is the region of the heart.

Aristotle consistently argued that the primary function of the wet and cold brain was to cool the warm blood coming from the heart.[3]

The most striking exception to this widespread neglect of the brain is the medical treatise, *On the Scared Disease*, written around 400 BCE. This short essay describes epileptic seizures in children, adults, and the old and their causes in entirely natural rather than in divine or magical terms. The author, who may have been Hippocrates, concludes that epilepsy provides proof that the brain controls mind and behavior:

> Men ought to know that from nothing else but the brain come joys, delights, laughter and sports, and sorrows, griefs, despondency, and lamentations. And by this, in an especial manner, we acquire wisdom and knowledge, and see and hear.

On the Sacred Disease is an isolated flash of insight within an ancient world that fails to recognize the brain as the seat of the soul. The foundational texts of Christianity, the Old and the New Testament, are no better; not a single reference to the brain; but a legion to the heart.

Cardiac-centric imagery and language are deeply engrained in our customs and language today—we love somebody with all our heart; we give our Valentines a heart-shaped chocolate rather than a hypothalamic-shaped sweet. There are hundreds of "Sacred Heart" churches and academies, but none dedicated to the "Sacred Brain." It wasn't until the closing decades of the seventeenth century that the heart was discovered, by gruesome live animal experiments, to be nothing more than a muscle, a biological piston, circulating blood throughout the body.

Some early anatomists knew that the brain was intimately tied to sensation and movement. The most influential was the second-century Common Era physician Galen, whose clinical knowledge derived from working at a gladiator school. Galen argued that the vital spirit that animates humans flows up from the liver to the heart and into the head. There, inside the ventricles, the brain's interconnected fluid-filled cavities, the vital spirit becomes purified into thought, sensation, and movement.

Galen's ideas dominated the next thousand years and found their apotheosis in the belief of the Church Fathers and Scholastic philosophers that the ventricles were the *sensorium commune*, where all senses combine to give rise to thought and action; the brain's gray matter—too mushy, coarse and cold to host the sublime soul—was merely pumping the vital spirit from the ventricles into the nerves. In a world with no mechanistic notions beyond those related to hydraulics, no idea about chemical metabolism and electricity, such explanations sounded at least vaguely plausible.

As the centuries accumulated, the main intellectual activities were eristic and sterile reinterpretations of classical authors and biblical exegesis (there is a reason this period is known as the Dark Ages). Medieval scholars were concerned with interior, spiritual matters while the systematic manipulation of nature, experimental philosophy, still lay far in the future.

Starting with the Renaissance and accelerating during the Enlightenment and the religious strife of the Reformation, attitudes shifted toward a more externalized, empirical view of the world. *Scientia* and *religion* diverged and began to denote different bodies of knowledge and research methodologies. Natural theology and natural philosophy emerged as precursors to modern science. The publication of *Cerebri anatome* in 1664 by the English doctor Thomas Willis heralded the onset of the brain-centric age with its meticulous drawings of the brain's convolutions that didn't just resemble intestines, as in traditional texts.[4]

It was a slow awakening to reason, though; until the early nineteenth century, sick people were at the brunt edge of bizarre medical treatments—constant bloodletting as a therapeutic and prophylactic for most any ailment, ingesting an amazing variety of concoctions of animal organs, strange plants, one's own urine, and so on. The nobler the patient, the worse the treatment—King Charles II of England was purged, cupped, and drained of quarts of his blood (by leeches and knifes) before he succumbed to kidney disease.

The early nineteenth century saw the rise of brain-based explanations.[5] Two trail-blazers were the German physician Franz Joseph Gall and his assistant Johann Spurzheim. Based on systematic dissection of human and animal cadavers, Gall formulated a thoroughly materialistic, empirically based account of the brain's gray matter as the sole organ of the mind. This organ was not homogeneous but an aggregate of distinct parts, each with its own distinct function; the functions themselves are scarcely recognizable today—constructiveness, acquisitiveness, secretiveness, gregariousness, benevolence, veneration, firmness, self-esteem, and reproductive love.

Using the shape and bumps of the skull, Gall and Spurzheim inferred the size and import of the organ underneath the cranium, and diagnosed the mental character of the individual examined. Their phrenological method proved immensely popular, as it appealed to the growing middle class as sophisticated and modern. Phrenology was used to classify criminals, lunatics, the eminent, and the infamous. As there is no discernible relationship between the shape of the external skull and the size and function of the underlying neural tissue, phrenology eventually lost favor as a reputable method.

The brain's building blocks, as with any other bodily organ, are its cells. This realization depended on the invention of specialized dyes to stain the extensive processes of individual cells in the second half of the nineteenth century. It was the Spanish anatomist Santiago Ramón y Cajal, the effective patron saint of neuroscience, who revealed neurons in all their stupendous glory. Just as there are kidney cells that are quite distinct from blood or heart cells, so there are different types of neurons and their nonneuronal partner cells, maybe as many as a thousand.[6] Today, his ink and pencil drawings of the neural circuits of the brain adorn museum exhibits, coffee table books, and my left bicep (as a tattoo).

Think of those National Geographic documentaries where a small plane captures the immensity of the Amazon by flying for hours over the jungle. There are as many trees in the rainforest as there are nerve cells in a single human brain. The vast morphological diversity of these trees, their distinct roots, branches, and leaves covered with vines and creeping crawlers, is comparable to that of nerve cells.

Ramón y Cajal put forth the *neuron doctrine*, the core dogma of neuroscience: the brain is a vast and tightly enmeshed lace of distinct cells, touching each other at specialized junctions known as synapses. Information flows in one direction, from thousands of synapses made onto the dendritic trees of neurons, their roots, to the cell body. From there the information is distributed via the neuron's single output wire, the axon, to thousands of other neurons in the next stage of processing. And so the circle closes—neurons endlessly talking to each other (some specialized neurons send their output to muscles). This silent conversation is the external manifestation of the subjective mind.

The physical basis of this neuronal banter is electrical activity. Each synapse[7] briefly increases or decreases the electrical conductance of the membrane. The resultant electrical charge is translated, via sophisticated, membrane-bound machinery in the dendrites and the cell body, into one or more all-or-none pulses, the fabled action potentials or spikes. They are about one-tenth of a volt in amplitude and last less than one-thousandth of a second. Spikes travel along the axon to the synapses and dendrites of the next set of neurons and the cycle recommences.

The final and decisive shift to a brain-centric view of life and death followed the invention of mechanical ventilators and cardiac pacemakers in the second half of the last century. Until then, everybody knew what death looked like—the lungs ceased breathing and the heart stopped beating. Today it's more complicated, as death has moved from the chest to the head—we are dead when our brain has irreversibly lost its function, even though the rest of the body may still be alive. I shall return to this morbid theme in chapter 9.

Conscious States and States of Consciousness

Before going any further, let me discuss a critical distinction between *conscious states* and *states of consciousness*, corresponding to the difference between the *transitive* (as in "conscious of pain") and the *intransitive* (as in "losing consciousness") usage of consciousness.

Watching the rays of the setting sun reflecting off distant mountains, lusting after someone, feeling schadenfreude at the comeuppance of a rival, or experiencing rising terror during a routine visit to a physician—each is a subjective experience with its own distinct coloratura. Our waking hours are filled with a never-ending stream of such conscious states or experiences whose content constantly changes. Maintaining the same content for more than a few seconds is challenging. Such stasis requires either a powerful stimulus—a loud alarm or a persistent migraine—or intense concentration—lying awake in a sleeping bag while tracking the sounds of something man-sized moving stealthily through a dark forest, being absorbed by mental arithmetic, obsessing about a recurring thought. But even here the minutiae of the content are ever shifting, trembling, never static. This most likely reflects the precarious equilibrium of the underlying neural assemblies.

The content of consciousness is fickle, continually changing within a fraction of a second. Like ripples and waves on the surface of a pond stirred by powerful currents beneath the surface—currents that represent the barely acknowledged ebb and flow of unconscious emotions, memories, desires, and fears. Like the distinct instruments of an orchestra that weave in and out.

All of this happens when we are awake and aroused in the physiological and psychological sense of being alert and ready to respond to a sound, sight, or touch. This is one state of consciousness.

When we sleep, consciousness fades. We spend one-quarter to one-third of our life asleep, more when we're young and less as we age. Sleep is defined by behavioral immobility (which is not absolute, as we continue to breathe, move our eyes, and occasionally twitch a limb) and reduced responsiveness to external stimuli. We share this need for daily sleep with all animals.

When awakened from sleep, especially early in the night, it seems as if we come to the light as from a limbo. We weren't there, and then, suddenly, we hear somebody calling our name and we become aware. From nothing into being. This is another state, characterized by not being conscious. Conversely, when spontaneously waking up in the morning, we can recall vivid sensorimotor experiences, often with an accompanying mundane or melodramatic narrative. We are magically transported to another realm where we run and fly, meet old lovers, children, faithful animal companions of yesteryear, all while our body is immobile, unresponsive, and largely disconnected from its environment. Dreaming, another state of consciousness, is an evocative feature of life that we take for granted.[8]

These three distinct states of consciousness are reflected in distinct electrical brain activities whose faint echoes can be picked up by electrodes on the scalp, the skin covering the skull. Just as the surface of the sea is in ceaseless commotion, so is the surface of the brain, reflecting the tiny electrical currents generated by cortical neurons.

The German psychiatrist Hans Berger pioneered electroencephalography (EEG) in his lifelong quest to prove the reality of telepathy. He first recorded the brain waves of a patient in 1924 but, filled with doubt, did not publish his findings until 1929. EEG became the foundational tool of an entire field of medicine, clinical neurophysiology, although Berger was never accorded any significant recognition in Nazi Germany and hanged himself in 1941, despite being nominated several times for the Nobel Prize.

The EEG measures the tiny voltage fluctuations (10 to 100 microvolts; figure 5.1) generated by electrical activity across neocortex, the brain's outer surface responsible for perception, action, memory, and thought. Different types of semiregularly occurring waves are named after their dominant frequency band. These include alpha waves in the 8–13 cycles per second or Hertz (Hz) range, gamma waves in the 40–200 Hz range, and delta waves in the 0.5–4 Hz band. Their irregular nature reflects activity in large coalitions of neurons with fluctuating membership. Yet the overall architecture and

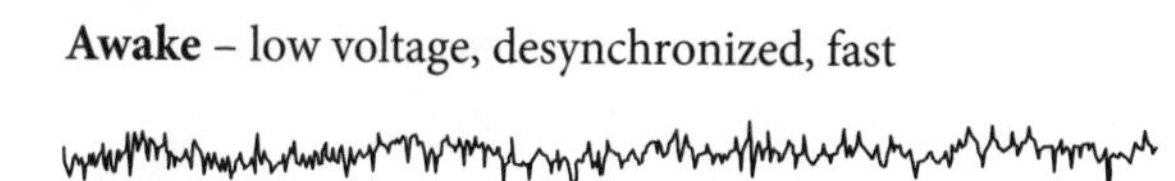

Drowsy – alpha waves

REM sleep – low voltage, desynchronized, fast with sawtooth waves

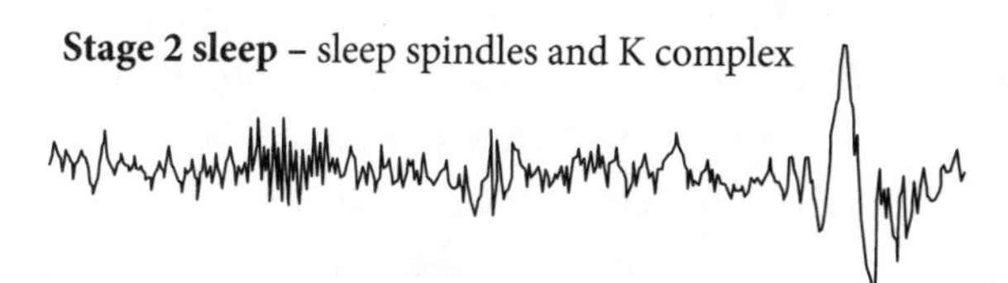

Deep sleep – large voltage, delta waves

Figure 5.1
Brain waves: Different brain states—alert, excited, drowsy, deeply asleep, dreaming—are reflected in distinct patterns of EEG activity measured by electrodes across the scalp. They are diagnostic of different states of consciousness in health and disease. (Redrawn from Dement & Vaughan, 1999.)

morphology of these waves, and their progress across the diurnal cycle and across lifespan, evolve in an orderly and lawful manner.

Clinical EEG setups can have as few as 4 or as many as 256 electrodes spread across the scalp. Using the EEG as an investigational tool, researchers were astounded to discover in 1953 that the sleeping brain transitions a couple of times every night between two different states—rapid eye movement (REM) sleep and deep or non-rapid eye movement sleep (non-REM).[9] REM sleep is characterized by low-voltage, choppy, rapidly changing brain waves, fast motion of the eyes and complete muscle paralysis. REM sleep is also called paradoxical sleep as parts of the brain are as active as during wakefulness. In contrast, deep or non-REM sleep is marked by slowly rising and falling electrical waves of larger amplitude. Indeed, the deeper and more restful the sleep, the slower and larger the waves that reflect the brain's idling, restorative activity. Today, consumer devices record our EEG via a slender band worn at night and use sounds that wax and wane in synchrony with deep-sleep waves to enhance sleep quality.[10]

For many decades, REM sleep was thought to be synonymous with dreaming (although we don't recall most of our dreams), and deep or non-REM sleep with a complete absence of any experience. This influential notion has been hard to dispel; but plenty of research proves it to be overly simplified. When subjects are woken at random and asked whether they experienced anything just prior to the awakening while their brain is monitored using a high-quality EEG setup, up to 70 percent deep sleep awakenings yield simple perceptual dream experiences. It is true that reported dreams upon REM awakening are more extended and complex, with elaborate life-like story lines and strong emotional overtones, than awakenings from deep sleep. Subjects recall no dream experiences at all in a significant minority of awakenings from REM sleep.[11]

Besides these three physiological states of consciousness (awake, REM, and deep sleep) that wax and wane with the diurnal cycle, societies throughout history have used, and abused, alcohol and drugs to modify mood, perception, stamina, and motor activity, entering altered states of consciousness. Of particular interest are the serotonin-receptor based hallucinogens and psychedelics—psilocybin, mescaline, DMT, tryptamines, ayahuasca, and LSD. Taken for spiritual and recreational purpose, these drugs change the quality and character of experience—inducing psychedelic colors, slowing down the perceived passage of time, losing one's sense of self.

Users speak of attaining a "higher" state of consciousness while tripping.[12] In *The Doors of Perception*, the book that kicked off the Age of Aquarius, Aldous Huxley describes one such episode:

> A moment later a clump of Red Hot Pokers, in full bloom, had exploded into my field of vision. So passionately alive that they seemed to be standing on the very brink of utterance, the flowers strained upwards into the blue.... I looked down at the leaves and discovered a cavernous intricacy of the most delicate green lights and shadows, pulsing with undecipherable mystery.

For clinical reasons, consciousness can be safely, rapidly, and reversibly turned off and on again for minutes or hours on end with a variety of agents. Anesthesia eliminates pain, distress, and haunting memories of surgery, an unnatural loss of consciousness whose mercies we take for granted. It is one of the great triumphs of modern civilization.

Pathological states of consciousness include coma and the vegetative state following gross trauma, a stroke, overdosing on drugs and/or alcohol, and so on. Here, consciousness has fled, yet parts of the victim's brain are still operating to support some housekeeping operations.

Any theory of consciousness will have to account for all of this massive data, on both conscious states and states of consciousness.

The Neural Correlates of Consciousness

In the late 1980s, as a young assistant professor at the California Institute of Technology in Southern California, I met with Francis Crick on a monthly basis. I was thrilled to have found a kindred soul willing to endlessly debate how the brain could produce consciousness. Crick was the physical chemist who, together with James Watson, discovered the double-helical structure of DNA, the molecule of heredity. In 1976, at the age of sixty, when Crick's interests shifted from molecular biology to neurobiology, he left the Old World to establish his new home in the New World, in La Jolla, California.

Despite our forty-year age difference, Crick and I struck up an intense teacher–student relationship. We worked closely for sixteen years and coauthored two dozen scientific papers and essays. Our collaboration continued literally until the day he died.[13]

When we started this labor of love, thinking seriously about consciousness was taken as a sign of cognitive decline and was ill-advised for a young scientist. But those attitudes changed. Together with a handful of philosophers and neuroscientists, we gave birth to a science of consciousness. It is no longer taboo, a field of study that shall not be named.

What takes places in the brain when experiencing the setting sun or a painful blister? Do some nerve cells vibrate at a magical frequency? Are there special consciousness neurons that turn on? Are they located in a particular region (shades of Descartes's pineal gland)? What is it about the biophysics of a chunk of highly excitable brain matter that links gray goo with the glorious surround-sound and technicolor that is the fabric of everyday experience? To answer these questions, Crick and I focused on an operational measure, the *neural* (or *neuronal*) *correlates of consciousness*, abbreviated in the literature as *NCC*. With help from David Chalmers's more rigorous formulation, it is defined as the *minimal neuronal mechanisms jointly sufficient for any one specific conscious percept* (fig. 5.2).[14]

Francis Crick and I meant the NCC language to be *ontologically neutral* (which is why we spoke of "correlates") with regard to the age-old battles of the -isms (dualism versus physicalism and their many variants; see chapter 14),

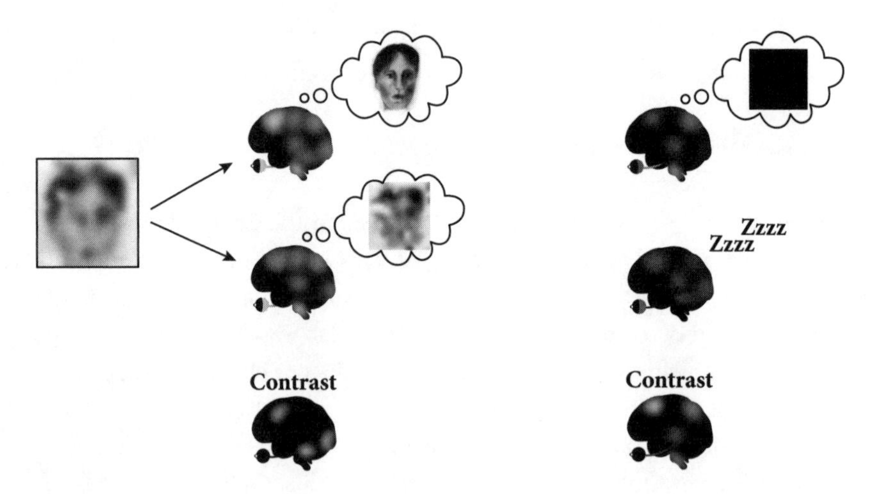

Figure 5.2

The neural correlates of consciousness: In the cartoon on the left, a flashed ambiguous image can be seen either as a face or as a nebulous black-and-white something. Contrasting the brain activity for these two percepts while you lie in a magnetic scanner identifies the *content-specific NCC* for the experience of seeing faces. A different experiment (on the right) compares brain activity when lying in a scanner with eyes closed to activity when you are deeply asleep. This pinpoints regions involved in the state of being conscious (the *full NCC*).

as we felt that at this point in time, science could not take a firm position with respect to resolving the mind–body problem. No matter what you believe about the mind, there is no doubt that it is intimately related to the brain.[15] The NCC is about where and how this intimacy takes place.

When defining the NCC, the qualifier "minimal" is important. For the brain as a whole can be considered an NCC: after all, the brain generates experience, day in and day out. But Crick and I were after the specific synapses, neurons, and circuits constitutive of experience.

Brain activity is utterly dependent on blood flow. Compress the left and right carotid arteries and consciousness ceases in seconds. There are no energy reserves to power the mind. The mind is fragile.

A sophisticated vascularization system delivers the life-sustaining blood to local neighborhoods within the brain. Trillions of disk-shaped red blood cells flow and tumble from the arteries into the capillary bed pervading all neural tissue, where the hemoglobin molecules inside these blood cells give up their precious cargo of oxygen to support cellular activity.[16] In the process, the cells change color from scarlet to dark red, before exiting neuronal tissue via the veins to pick up fresh oxygen in the lungs. Fortuitously for brain scientists, the response of hemoglobin to a magnetic field likewise changes when the oxygen is released, from slightly repulsive to weakly attractive. This effect is exploited by functional magnetic resonance imaging (fMRI), registering changes in oxygenation, blood flow, and volume, collectively called *hemodynamic response*. It is taken as a proxy for neural activity. That is, both blood flow and blood volume increase homeostatically in response to energetically expensive processes, such as active synapses and electrical spiking.

In a typical brain-imaging experiment, you lie inside a long, narrow cylinder, surrounded by heavy machinery (yes, it can be claustrophobic), while watching partially obscured pictures of faces flashing for 1/30th of a second onto a monitor. As explained in chapter 2, whenever you see a face, you push the "yes" button; when you see a chiaroscuro pattern of light and dark, you press "no." When images are visible only briefly, your brain often doesn't have enough time to form a coherent view; sometimes you see a face and sometimes you see something vague, uninterpretable, perhaps caused by the random twitching of a neuron. Your responses are sorted into two categories, "faces" and "non-faces," and the associated brain activities are compared (fig. 5.2). This contrast isolates those regions significantly more active when you see a face than when you do not—a set of regions in visual cortex, including the *fusiform face area* (FFA) on the underbelly of cortex, one on each side.[17] More generally, this procedure identifies a candidate region for a content-specific NCC, with the content being "the experience of seeing faces" in this instance.

Additional experiments are needed to determine that this activity relates to experience and not to pressing the "yes" button. Another confound is the task itself, requiring you to be mindful of the instructions and press the appropriate buttons. Ruling out that the FFA is involved in storing and following instructions (rather than seeing faces) can be done by a "no-task" paradigm. Other influences that must be investigated are the effects of selective visual attention, eye movements, and so on. These complications occupy the life of experimentalists and their graduate students and fill the literature.

In the late eighteenth century, the Italian physician Luigi Galvani discovered that electricity transmitted via nerve fibers made frog muscle twitch. Studying animal electricity gave birth to the science of electrophysiology. Galvani's successors discovered that electrical stimulation of the exposed brain made subjects jerk a limb, see lights, or hear sounds. By the middle of the twentieth century, electrical stimulation had become routine clinical practice.

Activating the NCC by such means should trigger the associated percept while suppressing the NCC should prevent the experience. Both predictions have been verified for faces and the fusiform face area. A study of epileptic patients conducted at Stanford University by the neurologist Josef Parvizi recorded the electrical signal from implanted electrodes to confirm that, indeed, both the left and the right fusiform areas responded selectively to faces. Parvizi then directly injected electrical current into the fusiform face area with the same electrodes (fig. 6.6). Stimulating the right fusiform area led one patient to exclaim: "You just turned into someone else. Your face metamorphosed. Your nose got saggy and went to the left."[18] Others reported similar distortions that bring to mind portraits painted by Francis Bacon. This didn't happen when nearby regions were stimulated or during sham trials when Parvizi pretended to inject current. For these patients, the fusiform face area seems to be an NCC for seeing faces,[19] as activity here correlates closely and systematically with seeing faces and stimulation of this region alters the perception of faces.

Furthermore, damage to this region can lead to *prosopagnosia* or face blindness. Affected individuals are unable to recognize familiar faces, including their own.[20] Faces of spouses, friends, celebrities, presidents all look alike, indistinguishable as pebbles in a riverbed. In its more severe forms, patients can't even recognize a face as a face anymore. They perceive the distinct elements making up a face, the eyes, nose, ears, and mouth, but they can't synthesize them into the unitary percept of a face. Intriguingly, these patients may still react unconsciously to familiar faces, with their autonomic system responding with an enhanced galvanic skin response. The unconscious has its own ways of detecting familiar faces.

Any change in the NCC will, of necessity, alter the character of the experience (including having none). However, if the background conditions change but the NCC does not, the experience will remain the same.

Related paradigms seek to identify the full NCC: the union of content-specific NCC for all possible experiences. This is the neural substrate that determines whether we are conscious of anything at all, irrespective of its specific content. One such experiment compares brain activity while lying quietly awake, with eyes closed, to activity when deeply asleep (as in fig. 5.2), not an easy thing to do in the loud and tight confines of a magnetic scanner. Again, a thousand complications loom; the devil is in the details.

From the edge of history until well into the seventeenth century, the heart was thought to be the seat of the soul. Today we know that it is the brain that is the substrate for the mental. So that is progress. But it doesn't stop there. In science's ceaseless drive to nail down mechanisms at their relevant level of causality, we need to probe further and ask which portion of this organ's three pounds of matter is most relevant to consciousness. I take this up next.

6 Tracking the Footprints of Consciousness

Let's roll up our sleeves and get down to the business of identifying which bits and pieces of brain matter are most closely tied up with consciousness. It turns out that many regions of the central nervous system can be dispensed with for experience. Bioelectrical activity in millions of neurons in disadvantaged neuro-neighborhoods does not contribute any one conscious feeling, while other regions are much more privileged. Where's the difference?

Consider the spinal cord, a long tube of nervous tissue inside the backbone. Roughly 18 inches long, it houses 200 million nerve cells.[1] If the spinal cord is completely severed by trauma in the neck region, victims are paralyzed in legs, arms, and torso; they lose control of their bowel, bladder, and other autonomic functions, and they lack bodily sensation. Confined to a wheelchair or bed, their situation is grim. Yet quadriplegics or tetraplegics continue to experience life in all its varieties—they see, hear, smell, feel emotions, imagine, and remember much as before the incident that irrevocably changed their life, refuting the myth that consciousness is an automatic by-product of nervous activity. More is needed.

The Brainstem Enables Consciousness

The spinal cord merges into the two-inch long brainstem at the base of the brain proper (fig. 6.1). The brainstem combines the functionality of a power plant with that of Grand Central Station. Its neural circuits regulate sleep and wakefulness, and the pulsation of the heart and lungs. Through its narrow confines pass most of the cranial cables innervating the face and the neck, incoming sensory (touch, vibration, temperature, pain) signals, and outgoing motor signals.

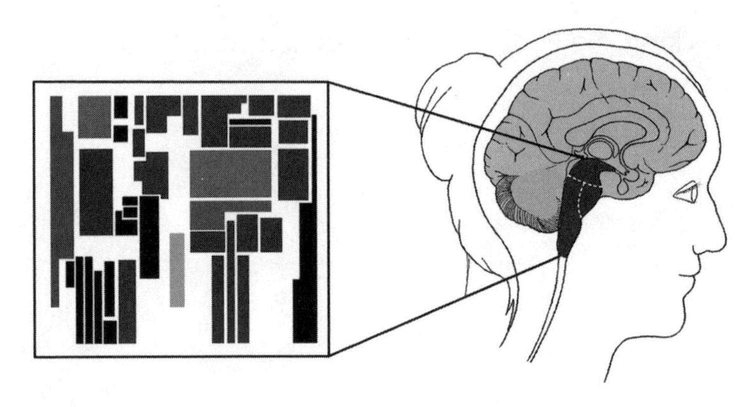

Figure 6.1
The brainstem as a background condition for consciousness: The brainstem reticular formation, spanning the medulla, pons, and midbrain (right), houses more than forty nuclei (left). Collectively, they regulate sleep and wakefulness, arousal, breathing and heart rate, temperature, eye movements, and other critical functions. Its neurons enable experience but do not provide the content for any one experience. The size of each rectangle reflects the relative size of each nucleus within the brainstem. (Redrawn from Parvizi & Damasio, 2001.)

If the brainstem is damaged or compressed, death frequently follows. Even quite focal destruction can lead to a profound and sustained loss of consciousness, especially if the damage occurs simultaneously on both left and right sides. This became apparent during the "sleeping sickness" pandemic (the *encephalitis lethargica*) emerging from the European battle fields of World War I.[2] It induced a profound, statue-like sleep in most of its victims, and hyper-arousal in others. The sleeping sickness killed an estimated one million people worldwide. The responsible culprit remains unidentified and presumably, at large. The neurologist Baron Constantin von Economo meticulously dissected the brains of its victims and discovered two discrete sites of infection in their brainstem, one in the hypothalamus promoting sleep and another one in the upper brainstem fostering wakefulness. Depending on which region was affected, victims were either hyper-somnolent or hyper-vigilant. Von Economo's discovery provided proof that sleep is not a passive state, brought on by the nightly loss of sensory stimulation and a tired body, but is a specific brain state regulated by a welter of circuits.

The brainstem houses at least forty distinct groups of neurons in cellular assemblies named the reticular formation or ascending reticular activating system. Each population uses its own neurotransmitter, such as glutamate, acetylcholine, serotonin, noradrenaline, GABA, histamine, adenosine, and orexin, which modulates, either directly or indirectly, the excitability of cortex and other forebrain structures. Collectively, they access and control signals relating to the internal milieu: breathing, thermal regulation, REM and non-REM sleep, sleep–wake transitions, eye muscles, and the musculoskeletal frame.[3]

Brainstem neurons enable consciousness by suffusing cortex with a cocktail of neuromodulatory substances, setting the stage on which mental life plays out. But do not confuse them with the actors that perform the play. The brainstem doesn't provide the content of any one experience. Patients with spared brainstem function but widespread cortical dysfunction typically remain in a behaviorally unresponsive state, without signs of consciousness of self or their environment.

Numerous processes must be in place to give rise to consciousness. Your lungs, like bellows, have to pull oxygen out of the air and deliver it to trillions of red blood cells that your heart pumps throughout the body and the brain. When the carotid arteries delivering oxygenated blood to the brain are blocked, you lose consciousness, fainting within seconds. Of course, blood flow by itself is insufficient for mind—a comatose patient whose heart is beating gives silent testimony to this. It is less appreciated that the mental likewise presupposes finely tuned brainstem circuitry, just as the electrical power supply is necessary for a laptop to work. The distinction between the necessary condition for any one specific experience to occur (the content-specific NCC) and those that enable the conscious state (the *background condition*) is difficult to discern in clinical practice, when an unconscious car accident victim is brought into the emergency room. But conceptually, the distinction is clear—the brainstem enables experience but is incapable of determining it.

Losing the Cerebellum Does Not Affect Consciousness

Where we do not find something can be as informative as where we do. This is spectacularly true of the cerebellum, the "little brain" tucked beneath the cortex, at the back of the head. The cerebellum instantiates the automatic feedback processes necessary to learn to coordinate the bodily senses and muscles needed for everyday living—standing, walking, running, using utensils, speaking, playing with toys, dribbling a ball, and so on. Acquiring and retaining these skills requires a never-ending dialogue between what was sensed by the eyes, the skin, the equilibrium organ in the inner ears, the stretch and position sensors in our muscles and joints, and so on, what the brain intends to do about this and what was actually executed by the body's musculoskeletal system.

The brain's most distinct neurons are cerebellar Purkinje cells (fig. 6.2) whose fan-shaped dendritic tree is the recipient of a staggering 200,000 synapses. Purkinje cells have complex intrinsic electrical responses and their axons convey the cerebellum's output to the rest of the brain. They are stacked, like books on a shelf, within the folds making up the cerebellar sheet. Collectively, Purkinje cells receive excitation from a mind-blowing 69 billion granule cells—four times more than all the neurons in the rest of the brain combined![4]

What happens to consciousness if portions of the cerebellum are lost to a stroke or to the surgeon's knife? I recently spoke at length with an eloquent young doctor. More than a year earlier, surgeons had removed an egg-sized chunk of his cerebellar tissue containing a glioblastoma, an aggressive brain

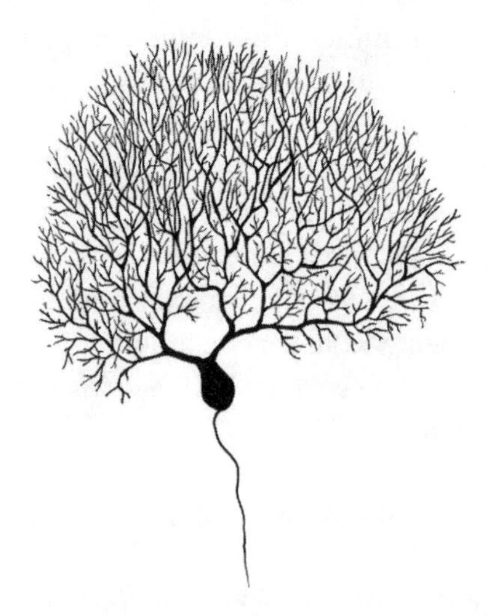

Figure 6.2

A Purkinje cell of the human cerebellum: Its striking coral-shaped dendritic tree receives input from several hundred thousand synapses. About ten million Purkinje cells provide the sole output of the cerebellum. Yet none of this circuitry generates conscious experience. (Redrawn from Piersol, 1913.)

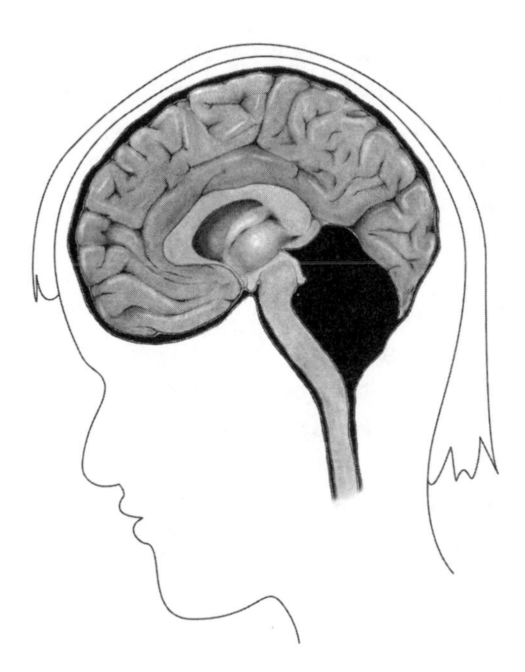

Figure 6.3
Living without a cerebellum: Drawing of a structural scan of a woman born with a gaping hole, filled with cerebrospinal fluid, where her cerebellum ought to be. Despite various motor deficits, she is conscious. (Adapted from Yu et al., 2014.)

tumor. Remarkably, though he lost his previous ability to play the piano effortlessly and type fluidly on his smartphone, he retained his conscious experience of the world, his ability to recall past events and project himself into the future. And this is typical. Some patients are not only clumsy but also have deficits in their thinking skills,[5] yet their subjective experience of the world remains intact.

More extreme is the case of a twenty-four-year old Chinese woman with mild mental impairment, slurred speech and medium motor deficits. During a brain scan, doctors discovered a cavern, filled with cerebral spinal fluid, where her cerebellum should have been (fig. 6.3). She is a rare instance of someone born without a cerebellum. Yet she leads a normal life with a young daughter, and experiences the world around her fully. She is no zombie.[6]

Purkinje cells are among the most elaborate of all neurons; the cerebellum maps the body and outside space onto its tens of billions of neurons. Yet none of this seems sufficient to generate consciousness. Why not?

Important hints can be found within its highly stereotyped, crystalline-like circuitry. First, the cerebellum is almost exclusively a *feedforward* circuit. That is, one set of neurons feeds the next one that, in turn, influences a third one. There are few recurrent synapses that amplify small responses or lead to tonic firing that outlasts the initial trigger. While there are no excitatory loops in the cerebellum, there is plenty of negative feedback to quench any sustained neuronal response. As a consequence, the cerebellum has no reverberatory, self-sustaining activity of the type seen in cortex. Second, the cerebellum is functionally divided into hundreds or more independent modules. Each one operates in parallel, with distinct, nonoverlapping inputs and output.

What matters for consciousness is not so much the individual neurons but the way they are wired together. A parallel and feedforward architecture is insufficient for consciousness, an important clue to which we will return.

Consciousness Resides in Cortex

The gray matter making up the celebrated cerebral cortex of each hemisphere, the brain's outer surface, is a laminated sheet of nervous tissue about the size, width, and weight of a fourteen-inch pizza with toppings (fig. 6.4). More than ten billion pyramidal neurons, with many recognized subtypes, provide its scaffolding, organized vertically, like trees in a forest, perpendicular to the cortical surface, intermixed with neurons that only form local connections, so-called interneurons. Pyramidal neurons are the workhorse of cortex, sending their outputs to other cortical sites, both near and far, including to the opposite cortical hemisphere. They also relay signals to the thalamus, claustrum, basal ganglia, and elsewhere. Intentions turn into actions by dint of specialized pyramidal neurons at the bottom of the cortical sheet that connect to motor structures in the brainstem and spinal cord.[7]

Collectively, legions of these axons bundle into fibers that anatomists call tracts, such as the commissural tract that connects the two hemispheres, or the corticospinal tract that conveys motor signals from cortex to the spinal cord. Tracks make up the white matter of the brain, whose light appearance is due to the fatty content of the myelin that surrounds axons and ensures high-speed conduction for the action potentials zipping along them.

Think of the fourteen-inch pizza that is cortical gray matter with its billions of cables like ultra-slender spaghettini, dangling from the bottom of

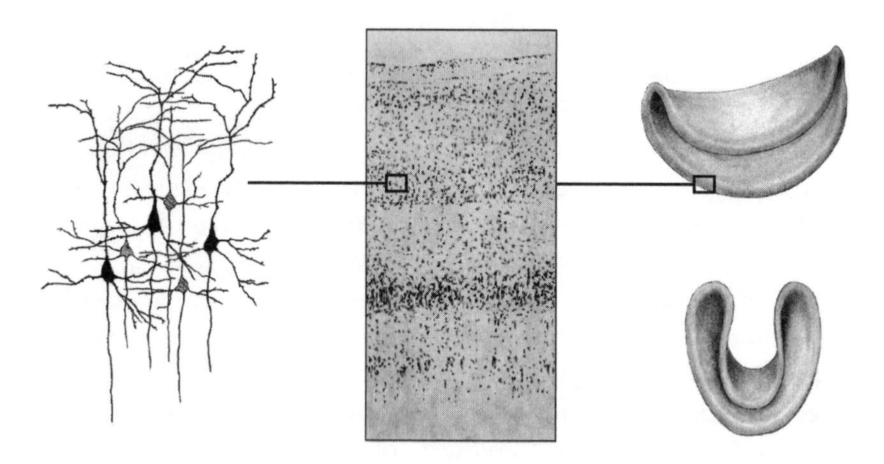

Figure 6.4
The neocortical sheet: Neocortex is a vast lace of pyramidal neurons and interneurons. The chiaroscuro pattern of cell bodies in the middle drawing, oriented from top to bottom, resolves into six layers that make up this structure. Shaped like a highly folded pizza or pancake, it forms the cortex's gray matter (right). The electrical activity of its neurons is the physical substrate of experience.

the cortical dough. Two of these highly folded sheets and their wiring are crammed into your skull.

Cortex is subdivided into neocortex, a defining hallmark of mammals, and the evolutionarily older archicortex, which includes the hippocampus. Zones within the highly organized neocortical tissue are most closely associated with subjective experience.

Losing Chunks of Posterior Cortex Causes Mind Blindness

Localized destruction in the brainstem can leave you in a stupor, coma, or worse; yet if you lose the function of small posterior (i.e., in the back) regions of your neocortex, you may still walk about, recall recent events, and behave appropriately but lack one or more categories of experiences.

This loss is not caused by a sensory defect in the eyes, ears, and other sense organs, nor by an inability to speak, or generalized mental deterioration such as dementia. A typical patient may not recognize keys on a chain dangling in front of her. She can see their shiny texture, sharp lines, and silvery metal, but her brain is unable to put these visual percepts together and recognize keys. Yet if she grasps them or if they are shaken so she can hear them jingle, she immediately calls out "keys." Such a deficit, called *agnosia* (Greek for "absence of knowledge"), can be caused by a circumscribed stroke in posterior regions of neocortex. It can wipe out an entire class of percepts or feelings. Face blindness, described in the last chapter, is an agnosia specific for faces. Let me highlight three further deficits involving loss of color perception (*achromatopsia*), loss of motion perception (*akinetopsia*), and loss of knowing about these deficits (*anosognosia*).[8]

Patient A.R. suffered a cerebral artery infarct that briefly blinded him. He recovered his sight but permanently lost color vision—not everywhere, though; only in the upper-left quadrant of his field of view, due to a pea-sized lesion in his right visual cortex. His only other difficulty was distinguishing forms—he couldn't read text—confined again to the upper-left quadrant.

Not uncommon for a cortical-induced deficit, A.R. did not know that part of his world was colorless. How can this be? If a portion of a computer monitor only displays black pixels while the rest of the screen remains normal, you would immediately notice. So how come A.R. didn't? Because his situation is different from yours. As the color centers in your brain are functional, you see a black region and know implicitly that it is not red, green, or blue. Given that A.R.'s very apparatus that determines color was damaged, he didn't know what colors were (except in an abstract sense). Denying an objective sensory or motor deficit due to neurological damage is a form of agnosia termed *anosognosia*. It is really a deficit in self-awareness: not knowing what it is that one no longer knows.

Consider another patient suffering from a bilateral posterior cerebral artery stroke. Unable to read sentences or individual words, he failed to notice achromatopsia in his upper visual field. Yet he retained semantic knowledge of the colors of objects. Confronted by unambiguous evidence for his loss of color discrimination, he reluctantly admitted that he saw everything in shades of gray without having been aware of it. Remarkably, he wasn't perturbed by eating colorless food, explaining, "No, not at all! You just know what color your food is. Spinach, for example, is just green."

Over the following months, he did better on color tests but began to realize that the colors he did see looked grayish and dirty. That is, as the patient's color perception partially returned, so did his ability to know what colors he didn't perceive. This suggests that the same region generates both the conscious percept of color and the knowledge of what color is. This distinction is important: watching a black-and-white movie with intact cortex that signals the absence of color is quite different from being unable to see the colors in a movie because of cortical achromatopsia.[9]

Much rarer and more devastating is motion blindness. One of only a handful of known patients with this syndrome is L.M. She lost part of her occipital-parietal cortex on both sides to a vascular disorder and could no longer see motion. She had to infer that a car had moved by comparing its relative position over time. She retained normal color, space, and form perception. She lived in a world not unlike a nightclub, in which the dancers lit by strobe lights appear frozen, devoid of motion; or as if she were watching a greatly slowed-down movie in which the vivid sense of motion is lost.

What the study of such patients has taught me is that a broad zone in the temporo-parietal-occipital region of the posterior cerebral cortex, in the back, is the best current candidate for the NCC for sensory experiences. The actual substrate is, in all likelihood, subsets of pyramidal neurons within this *hot zone* that support specific phenomenological distinctions, such as for seeing, hearing, or feeling. This explains why losing circumcised chunks of neocortex in the back leaches colors from the world, renders faces meaningless, and eliminates the sense of motion. Some researchers, however, dispute this conclusion and localize the epicenter of experience toward more frontal regions. This question will need to be resolved experimentally.[10]

Is Prefrontal Cortex Necessary for Experience?

To save a patient's life from brain tumors or to ameliorate the effect of nervous storms—epileptic seizures—neurosurgeons cut, coagulate, or resect brain tissue.[11] Removing primary visual, auditory, or motor cortex has the expected effect: patients become partially or completely blind, deaf, or paralyzed. Removing tissue from the left temporal or left inferior frontal gyrus can leave patients unable to read (alexia), understand, and/or articulate speech (aphasia). Clinicians refer to these regions as "eloquent" cortex (fig. 6.5). In striking contrast is the vast expanse of cortical tissue in front of the premotor strip, known as prefrontal cortex, from which strikingly fewer reports can be elicited from patients. The function of this tissue is less obvious, as much of it is silent when stimulated.[12]

As the border between eloquent and non-eloquent cortex varies from patient to patient, it needs to be carefully mapped prior to surgical

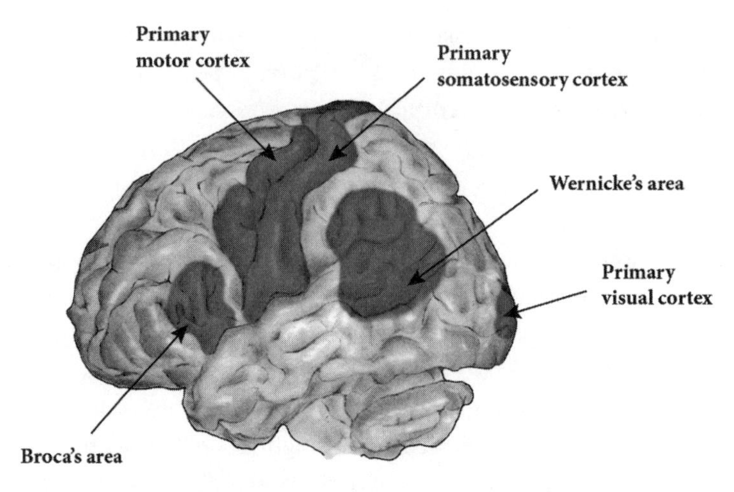

Figure 6.5
Eloquent cortex: Resecting primary sensory or motor cortices on either side, Broca's area in the left inferior frontal gyrus, or Wernicke's area in the left superior temporal gyrus causes permanent sensory, motor, or linguistic deficits. These regions are collectively known as *eloquent cortex*. Less appreciated is the converse—that large regions of prefrontal cortex can be surgically removed without apparent ill effect on conscious experience.

intervention. After the surgeon drills holes or otherwise trepans the skull, anesthesia is suspended (past its covering membranes, the brain has no pain receptors) so that the patient is awake and can articulate the effects, if any, of electrical stimulation. The neurosurgeon thereby delimits eloquent regions in the winding folds and bends of the cortical landscape.

The removal of silent prefrontal cortex causes no obvious sensory or motor deficit. Patients may not complain and their families may not notice any drastic deficit or problems. Impairments tend to be subtle and affect the higher mental faculties—a reduced ability to introspect, to regulate emotions, to spontaneously initiate behaviors, apathy, a lack of curiosity about the world. What is remarkable is how unremarkable these patients appear. The direct effects of posterior lesions on patients' mental life compared to the indirect effects of prefrontal lesions is well known among clinicians but is often unacknowledged by cognitive neuroscientists, a compelling illustration of what Oliver Sacks refers to as scotomas in the history of science—forgetting and neglecting inconvenient truths.[13]

Consider a famous patient: Joe A., a stockbroker. Because of his massive meningioma, the surgeon amputated almost the entirety of his frontal lobes. This radical lobectomy removed an astounding 230 grams of prefrontal tissue. Subsequently, A. acted childlike, distracted, exuberant, and boastful, lacked social inhibition, and so on. Yet he never complained of being deaf, blind or unable to remember. Indeed, the attendant neurologist commented:

> Yet, one of the salient traits of A.'s case was his ability to pass as an ordinary person under casual circumstances, as when he toured the Neurological Institute in a party of five, two of whom were distinguished neurologists, and none of them noticed anything unusual until their attention was especially called to A. after the passage of more than an hour. In particular, A.'s impairment of intellectual performance as such was never conspicuous on casual examination.[14]

In another patient, the surgeon resected one third of the anterior portion of both frontal lobes to eliminate debilitating epileptic seizures; the patient showed "a striking post-operative improvement in personality and intellectual capacity."[15] Both patients went on to live for many years, with no documented evidence that the removal of so much frontal tissue significantly affected their conscious sensory experience.

Brain scanners have revolutionized diagnosis, making such massive surgical interventions rare.[16] But accidents and viral infections persist. In a more recent tragedy, a young man fell onto an iron rod that completely penetrated both of his frontal lobes. Nevertheless, he went on to lead a stable family life—getting married and raising two children. Although displaying many traits typical of frontal lobe patients (e.g., disinhibition), he did not lose conscious experiences.[17]

Though prefrontal cortex is not needed to see, hear, or feel, does not imply that it has nothing to contribute to consciousness. In particular, judging one's own confidence in having seen or heard something—metacognition, or "knowing about knowing" (recall the four-point confidence scale in chapter 2)—is linked to anterior regions of prefrontal cortex.[18] Yet most of your daily experiences are more sensorimotor in nature—biking in dense urban traffic, listening to music, watching a movie, fantasizing about sex, dreaming. These are generated by the posterior hot zone.

The inference that prefrontal cortex is not critical for many forms of consciousness, based on lesion and stimulation data, illustrates the danger of relying too much on correlative evidence from neuroimaging to identify the NCC: because of the intricate connectivity of the brain, activity in the prefrontal cortex, basal ganglia, and even cerebellum may vary systematically with experience without being responsible for it.[19]

The importance of the posterior cortex to experience is compatible with Crick's and my speculation on the *unconscious homunculus* outlined in chapter 4. Prefrontal cortex and its closely connected regions in the basal ganglia uphold intelligence, insight, creativity; these regions are needed for planning, monitoring and executing tasks, and regulating emotions. The associated brain activities are by and large unconscious. Prefrontal regions act like a homunculus who receives massive inputs from the posterior cortex, makes decisions, and informs the relevant motor stages.

Much evidence supports the hypothesis that primary sensory cortices, defined as regions directly receiving sensory input via the thalamus, are not content-specific NCC. This has been most extensively studied for primary visual cortex at the back of the brain, the ultimate terminus of the output of the eye via a relay in the visual thalamus. While activity in primary visual cortex reflects information streaming up from the retina, it differs in striking ways from what you actually see. Your view of the visual world is provided by regions higher up in the cortical hierarchy. Similar conclusions have been reached for primary auditory cortex and primary somatosensory cortex.[20]

So if it is true that neither primary sensory cortices nor prefrontal cortex contributes to consciousness but the posterior hot zone does, where is the critical difference? Why are certain cortical neighborhoods privileged with respect to experience? I suspect it is the way they are interconnected.

Cortex in the back is topographically organized, with a grid-like connectivity, reflecting the geometry of sensory space. Conversely, the front appears more like a network with random connectivity, enabling arbitrary associations. This connectivity among cortical neurons in the back versus the front might make all the difference when it comes to consciousness (as I will discuss in chapters 8 and 13). Careful anatomical analysis will need to confirm this at the microcircuitry level.[21]

Electrical Brain Stimulation Elicits Conscious Experiences in the Back of Cortex

Electrically stimulating the occipital, parietal, and temporal lobes can trigger a litany of sensations and feelings—the face distortions discussed earlier, seeing flashes of lights (also called *phosphenes*), geometric shapes, colors and movement, hearing sounds, feelings of familiarity (déjà vu) or unreality, the urge to move a limb without actually moving it, and so on (fig. 6.6). The evoked visual percepts can be lawfully related to the underlying neural

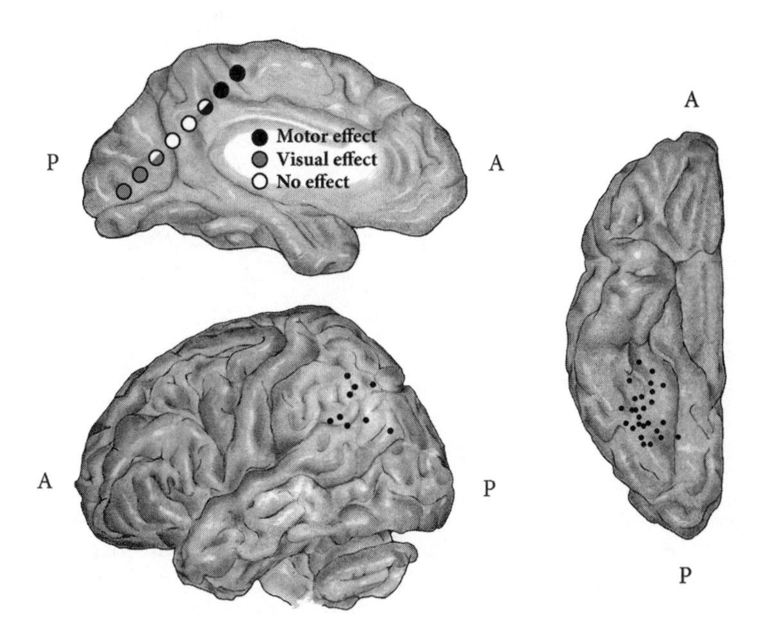

Figure 6.6
The geography of the posterior hot zone: Electrically stimulating the back of cortex can reliably trigger conscious sensations, demonstrated here in three exemplar studies. The drawings depict the left hemisphere as seen from three different directions, with P (for posterior) indicating the back of the brain and A (for anterior) its front. The medial view (upper-left) highlights stimulation that elicits visual percepts in occipital regions and motor responses once past the cingulate sulcus. The lateral view (lower-left) marks locations in posterior parietal cortex that trigger a conscious desire to act or move specific limbs. Stimulating sites within the fusiform gyrus on the bottom of cortex (right) causes distortions when seeing faces. (Upper-left image redrawn from Foster & Parvizi, 2017; lower-left image from Desmurget et al., 2009; image on right from Rangarajan et al., 2014.)

response patterns, to the extent that brain–machine interfaces in primary visual cortex are being considered as prosthetic devices for the visual blind. Electrical stimulation is compelling testimony to the intimate relationship between posterior cortex and sensory experiences.[22]

The neurosurgeon Wilder Penfield at the Montreal Neurological Institute assembled a vast catalog of information on the local topography of brain function from open skull neurosurgeries on patients treated for severe epileptic seizures. In one famous study, Penfield amassed data concerning *experiential responses*, vivid experiences or hallucinations—previously seen or heard, frequently familiar voices, music, or sights—from electrical

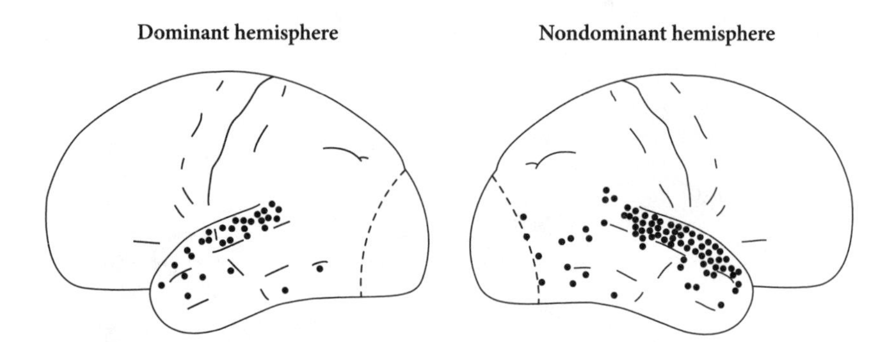

Figure 6.7
Experiential responses from the temporal lobe: Locations on the cortical surface where electrical stimulation by the neurosurgeon Wilder Penfield evoked experiences. There is a clear bias for the nondominant right hemisphere. Most of these complex visual and auditory hallucinations are triggered from the superior temporal gyrus; some can be triggered from other temporal regions or from parieto-occipital regions. Stimulation of the rest of the cortical windings did not evoke experiential responses in more than five hundred patients. (Redrawn from Penfield & Perot, 1963.)

stimulation of more than a thousand patients. Such responses are only found in posterior cortex, most often in the temporal lobe, and also at times from parieto-occipital regions (fig. 6.7).[23]

Stimulating the front of cortex is a different story. Far fewer clinical trials reliably elicit experiences, either simple perceptual ones or more complex experiential ones. Outside the primary motor and premotor regions, stimulation of which initiates behaviors in the appropriate muscle groups, and Broca's area in the left inferior frontal gyrus that interferes with language, tickling anterior gray matter elicits little sensation.[24]

Stories from *The Thousand and One Nights* tell of a magical brass lamp that, when rubbed, releases a powerful spirit, a djinn. Something similar occurs with the posterior hot zone. Zap it and the genius of consciousness appears. Conversely, prefrontal cortex is more of a fake brass lamp. To elicit experience, you'll have to find the few true spots to rub.

While cortex grabs the lion's share of headlines, other structures may also play an important role in the expression of consciousness. Francis Crick was fascinated, literally to his dying day, with a mysterious thin layer of neurons underneath the cortex called the claustrum. Claustrum neurons project to every region of cortex and also receive input from every cortical region. Crick and I speculated that the claustrum acts as the conductor of the cortical symphony, coordinating responses across the cortical sheet in a way that is essential to any conscious experience. Laborious but stunning reconstructions of the axonal wiring of individual nerve cells (which I call "crown of thorns" neurons) from the claustrum of the mouse confirm that these cells project massively throughout much of the cortical mantle.[25]

Neuroscientists won't be satisfied with pointing to a swath of gray matter as exuding consciousness; they want to dig deeper. Any piece of nervous tissue is a dazzling tapestry of tightly knotted neurons of various types, more finely woven that any Persian carpet from Isfahan. A quinoa-grain-sized piece of cortex contains 50,000 to 100,000 neurons of a hundred different types, a few miles of axonal wiring, and a billion synapses.[26] Where among these tangled networks are the specific agents responsible for any one conscious experience? Which neuronal cell types are the principals and which ones are bit players?

Going deep is almost always impossible in humans and requires experiments in mice and monkeys trained to respond to one or another percept while the activity of individual neurons is tracked and manipulated. This involves either mechanical silicon probes thinner than a single hair fiber or specialized microscopes that pick up the optical signatures of neurons engineered to turn their electrical pulses into flashes of green light.

We don't know the minimal number of neurons that constitute any one experience. Given that so many regions of the brain are out as NCC (the spinal cord, the cerebellum, primary sensory cortical regions and much of prefrontal cortex), I would guess that the fraction of neurons involved are a few percent or less of the brain's 86 billion neurons.

We might even have to hunt for the NCC down at the subcellular level, seeking mechanisms operating inside cells rather than across large neural coalitions, as is widely assumed. Indeed, some have hypothesized, as a possible NCC, all-or-none electrical events occurring in the dendritic tree of cortical neurons, a sort of handshake confirming that a bottom-up signal has encountered top-down feedback within a certain time window.[27]

Quantum Mechanics and Consciousness

Much ink has been spilled over arguments that quantum mechanics (QM) is the secret to consciousness. These speculations originate with the famous measurement problem. Let me explain.

Quantum mechanics is the established text-book theory of molecules, atoms, electrons, and photons at low energies. Much of the technological infrastructure of modern life exploits its properties, from transistors and lasers to magnetic resonance scanners and computers. QM is one of humanity's supreme intellectual achievements, explaining a range of phenomena that cannot be understood within a classical context: light or small objects can behave like a wave or like a particle depending on the experimental setup (*wave–particle duality*); the position and the momentum of an object cannot both be simultaneously determined with perfect accuracy (*Heisenberg's uncertainty principle*); and the quantum states of two or more objects can be highly correlated even though they are very far apart, violating our intuition about locality (*quantum entanglement*).

The most famous example of a prediction of QM at odds with common sense is Schrödinger's cat. In this thought experiment, an unfortunate feline is locked in a box, together with a diabolical apparatus that triggers a lethal gas upon the decay of a radioactive atom, a quantum event. There are two possible outcomes. Either the atom decays, thereby poisoning the cat, or it does not, in which the cat remains alive. According to QM, the box exists in a superposition of dead and live cat simultaneously, with the associated wave function describing the probabilities of these states. Only when a measurement is made, that is, only when somebody looks inside the box, does the system abruptly change from the superposition of two states to a single one. The wave function is said to "collapse" and the observer sees either a dead or alive cat.

That a conscious observer is required to convert the superposition of states of a quantum system into the single observable outcome has always troubled physicists.[28] If QM is really a fundamental theory of reality, it shouldn't need to invoke conscious brains and measuring devices. Instead these macroscopic objects should emerge naturally from the theory. Many solutions have been proposed, but none have found acceptance. It is this dilemma at the root of such a successful theory of reality that has prompted Roger Penrose, a brilliant cosmologist, to propose a quantum gravity theory of consciousness that remains popular with the public.[29]

Precious little evidence supports the idea that the brain exploits macroscopic quantum mechanical effects. Of course, the brain has to obey QM, as when photons of light meet retinal molecules inside photoreceptors. Yet the body's wet and warm operating regime is inimical to quantum coherency and superposition across neurons. Today's quantum computer prototypes need extreme vacuum and temperatures close to absolute zero to avoid decoherence, when so-called quantum bits disentangle and become regular bits of classical information theory. That's why quantum computers are so difficult to build.

Furthermore, it has never been properly explained why phenomenological aspects of consciousness or its neurobiological substrate require quantum properties.

I see no need to invoke exotic physics to understand consciousness. It is likely that a knowledge of biochemistry and classical electrodynamics is sufficient to understand how electrical activity across vast coalitions of neocortical neurons constitute any one experience. But as scientist, I keep an open mind. Any mechanism not violating physics might be exploited by natural selection.[30]

The previous chapters have covered both sides of the mind–body problem: the nature of experience and the bodily organ principally linked to it. The next two chapters describe integrated information theory and how it links these two seemingly disparate domains of life, the mental and the physical, in a rigorous manner. I will start off by sermonizing on the need for a theory, as this is not obvious to everybody, and what precisely it needs to explain.

7 Why We Need a Theory of Consciousness

The previous chapter could easily be expanded into an entire textbook, as knowledge of the neuronal footprints of consciousness is vast. Some parts of the nervous system, such as the spinal cord, the cerebellum, and most if not all of prefrontal cortex, are clearly out, while others, such as the posterior hot zone, are in. Biology is all about amazing molecular machinery and gadgets. It will be no different for consciousness. In the fullness of time, science will determine the events at the relevant level of granularity that constitute any one experience.

Once science sees the neural correlate of consciousness face to face, what then? It will be a momentous occasion to celebrate, with Nobel Prizes given out, editorials in newspapers, and textbooks galore. An abundance of drugs and medical treatments will flow from discovering this true NCC, ameliorating the myriad of neurological and psychiatric conditions our brains are prey to.

But we would still not understand at a conceptual level why *this* mechanism but not *that* one constitutes a particular experience. How can the mental be squeezed out of the physical? To paraphrase the New Testament and the philosopher Colin McGinn, how is the water of the brain turned into the wine of experience?

One of the most famous arguments against a materialistic conception of consciousness was formulated three hundred years ago by the German rationalist, engineer, and polymath Gottfried Wilhelm Leibniz, who invented calculus and binary numbers and built the first general digital calculator. He presented what is known as the *mill thought experiment*:

> Even if we had eyes as penetrating as you like, so as to see the smallest parts of the structure of bodies, I do not see that we would thereby be any further forward. We

> would find the origin of perception there as little as we would find it in a watch, where the constituent parts of the machine are all visible, or in a mill, where one can even walk around among the wheels. For the difference between a mill and a more refined machine is only a matter of greater and less. We can understand that a machine could produce the most wonderful things in the world, but never that it might perceive them.[1]

That is the gauntlet thrown down to materialism and its modern variant, physicalism. Zooming into the posterior hot zone with an electron microscope, we can see only membranes, synapses, mitochondria and other organelles (fig. 7.1). Were we to go even deeper with an atomic force microscope, individual molecules and atoms would come into focus. But never any experience. Where does it hide? Explaining this conundrum demands a fundamental theory of consciousness that we will turn to now.

Figure 7.1
Leibniz's mill argument updated to the twenty-first century: Over three hundred years ago, Leibniz pointed out that no matter how closely we zoom in to the body—epitomized by the most advanced technology of his day, windmills—we will never find experience but only levers, gears, shafts, and other mechanisms. Using today's instruments, we can indeed peer at the smallest organelles of the brain, the synapses (arrows), here in an electron-microscopic image of cerebral cortex. Where does experience hide? The bar indicates 1/1000th of a millimeter.

Integrated Information Theory

My experience is the starting point, my omphalos, from which I must abduce everything else, including the existence of an external world. That is, to understand how consciousness relates to the world at large—to dogs and trees, people and stars, atoms and the void—I must start with my own experience. This is the central insight of Saint Augustine (*Si fallor, sum*) and, over a thousand years later, of Descartes (*Cogito, ergo sum*).

The security of this navel, the only world I am directly acquainted with, allows me to capture the true character of every experience in a quintuplet of properties that are immediate and indisputable. These constitute the axiomatic bedrock from which I proceed onward, abducing the requirements for any substrate to instantiate an experience. This step yields the physical substrate of consciousness. For creatures like you and me, this substrate is identical to the neural correlates of consciousness (NCC) at the relevant level of spatiotemporal granularity.

Only by starting with raw experience can Leibniz's mill argument and its modern variants be successfully countered. The Australian-American philosopher David Chalmers coined the term "the hard problem" through a closely argued chain of reasoning involving zombies—imaginary creatures that look like us except that they have no sensations. Unlike their Hollywood namesakes, philosophical zombies act like regular folks; they have neither superpowers nor a predilection for human flesh. To lull us into a sense of complacency, zombies even speak about their feelings. But it's all a deep fake. Unfortunately, there is no way to tell a zombie from you and me.

Chalmers asks whether the existence of zombies is at odds with any natural law. That is, can we conceive of a world without experience yet still obeying the same physics as our world? The short answer is yes. None of the foundational equations of quantum mechanics or the theory of relativity mentions experience; neither does chemistry or molecular biology. In a book-length argument, Chalmers concludes that no natural law precludes the existence of such zombies. Put differently, conscious experience is an additional fact above and beyond contemporary science. Something else is needed to explain experience. He acknowledges the existence of bridging principles—empirical observations that link the material world to the phenomenal one (such as the observation that the brain has an intimate relationship to consciousness). But *why* certain bits and pieces of matter should have this close relationship to experience is a mystery that is hard, and perhaps even impossibly hard, to solve.[2]

Integrated information theory (IIT) does not bang its head against this cement wall; it doesn't try to squeeze the juice of consciousness out of the brain.[3] Rather, it starts with experience and asks how matter must be organized to support the mental. Does any sort of matter suffice? Are complex systems of matter more likely to host experience than less complex systems? What precisely is meant by "complex"? Is there a bias for organic chemistry over doped semiconductors? Or for evolved creatures over engineered artifacts?

IIT is a fundamental theory that links ontology, the study of the nature of being, and phenomenology, the study of how things appear, to the realm of physics and biology. The theory precisely defines both the quality and the quantity of any one conscious experience and how it relates to its underlying mechanism.

This theoretical edifice is the singular intellectual creation of Giulio Tononi, a brilliant, sometimes cryptic, polyglot and polymath renaissance scholar, a scientist-physician of the first rank. Giulio is the living embodiment of the *Magister Ludi* of Hermann Hesse's novel *The Glass Bead Game*, the head of an austere order of monks-intellectuals, dedicated to teaching and playing the eponymous glass bead game, capable of generating a near infinity of patterns, a synthesis of all arts and sciences.

IIT is a deep theory in the sense that it explains many facts, predicts new phenomena, and can be extrapolated in surprising ways that will be discussed in the remainder of the book. As more philosophers, mathematicians, physicists, biologists, psychologists, neurologists, and computer scientists become interested in IIT, we are learning more about its mathematical foundations and its implications for understanding existence and causality, for measuring physiological signatures of consciousness, and for the possibility of sentient machines.

From Phenomenology to Mechanisms

The opening chapter distilled five essential phenomenological properties out of the feeling of life. These are common to all, even the most rarefied experience: any experience exists for itself, is structured, is the specific way it is, is one, and is definite.

To bring the point home that these properties cannot be doubted, let us consider negating them. What would an *extrinsic* experience be? When it is the experience of somebody else, perhaps? But then it is not yours. How could an experience be unstructured? Even a content-less pure experience is still structured, as the whole is a subset of the whole, though not a proper one. What would it mean for an experience to be uninformative or generic? It is the way it is precisely because it is something—yellow, ice-cold, or stinky. Could an experience be more than one? That makes no sense, for your mind can't independently have two discrete experiences, next to each other. Finally, how could an experience be indefinite? If you look at the world, you see all of it; you don't see half of it, superimposed onto the whole and maybe a third experience of your dog sneaking into your field of view.

IIT starts with these five phenomenological properties, adopting them as axioms. In geometry or mathematical logic, axioms are foundational statements that serve as a starting point for deducing further valid geometric or logical properties and expressions. Just like mathematical axioms, these five phenomenological axioms—every experience exists for itself, and is structured, informative, one, and definite—don't contradict each other (*consistency*); none can be derived from one or more of the others (*independence*); and they are complete. From these premises, IIT infers the type of physical mechanisms necessary to support these five properties.

Each axiom has an associated postulate, a bridging principle that the system under consideration has to obey. Five postulates—intrinsic existence, composition, information, integration, and exclusion—parallel the five phenomenological axioms. These postulates can be thought of as abductions, inferences to the best explanation, from the axioms, in the sense I explained in chapter 2.[4]

Consider a physical system of connected nerve cells or an electronic circuit. Either one is a complicated set of mechanisms in a particular state. By mechanism, I mean anything that exerts a causal effect on other mechanisms. Anything that can make something else happen, perhaps only sometimes or only in conjunction with other entities, is a mechanism. Old-fashioned machines, such as windmills with wheels, levers, ratchets, and gears, to grind flour (fig. 7.1), are examples of mechanisms; indeed, the word derives from the Greek *mekhane*, "machine, instrument, or device." Neurons firing action potentials affecting all downstream cells they are wired to are another type of mechanism, as are electronic circuits, made of transistors, capacitances, resistances, and wires.

Consciousness needs some sort of mechanism. At one of my meetings with the Dalai Lama and Tibetan monks, the topic eventually turned to the Buddhist belief in reincarnation and, in particular, to the question of where the mind—with its memories—resides between consecutive incarnations. I answered by raising four fingers and counting down, one for each word—*No brain, never mind.* What I meant by this koan is that I can't conceive of consciousness existing in physical limbo—it requires a substrate. Maybe an esoteric one such as electromagnetic fields as in Fred Hoyle's sentient cloud of gas in *The Black Cloud.* But there has to be something. In the absence of anything, there can't be any experience either.

Given some substrate in some state, IIT computes the associated *integrated information* to determine whether that system feels like something, for only a system with a non-zero maximum of integrated information is conscious. Take a brain with some neurons that are, at that particular point in time, ON (spiking) while other neurons are OFF (not spiking); or a microprocessor with some transistors in an ON state, which means that their gate is storing an electrical charge that modifies the current in the underlying channel, while others are OFF, that is, their channels are nonconductive. As the circuit evolves, moving through different states, its integrated information changes, sometimes dramatically. Such variations in consciousness occur every day and every night in all of us.

Why Should Integrated Information Be Experienced?

Before I come to the mathematical innards of the theory, let me address one general objection to IIT that I frequently encounter. It runs along the following lines. Even if everything about IIT is correct, why should it feel like anything to have a maximum of integrated information? Why should a system that instantiates the five essential properties of consciousness—intrinsic existence, composition, information, integration, and exclusion—form a conscious experience? IIT might correctly describe aspects of systems that support consciousness. But, at least in principle, skeptics might be able to imagine a system that has all these properties but which still doesn't feeling like anything.

I answer this conceivability argument in the following manner. By construction, these five properties fully delimit any experience. Nothing else is left out. What people mean by subjective feelings is precisely described by these five axioms. Any additional "feeling" axiom is superfluous. Is there a mathematically unassailable proof that satisfying those five axioms is equivalent to feeling like something? Not to my knowledge. But I'm a scientist, concerned with the universe I find myself in, and not with logical necessity. And in this universe, so I argue in this book, any system that obeys these five axioms is conscious.

The situation IIT finds itself in is not dissimilar to the position of modern physics. Quantum mechanics is by far the best description of what exists at the microscale. Can it be proven that quantum mechanics has to hold in the universe? No. One can certainly conceive of universes with different microphysical laws than those that govern our universe (for instance, those of classical physics). Or consider the problem of *fine-tuning*: some of the numbers that make their appearance in cosmology and in particle physics are either very, very small or very, very large. The equations with their parameters explain the data very well but not why these numerical values hold. Who or what tuned them to their precise value? Physicists have converged on several broad classes of answers.

The least popular one is that there is no deeper explanation, at least none accessible to the human mind. These equations with their settings explain the observed world and that's the way it is. A brute fact; end of story! A second class of explanations posits that this universe, along with its particular laws, was created; either by a Supreme Being, as in traditional religions, or by some alien civilization with god-like powers. The third broad explanation is the *multiverse*, mentioned earlier. This is the very large set of universes that make up the cosmos, each one following different laws.[5] We happen to live in the one that obeys quantum mechanics, is conducive to life, and in which integrated information gives rise to experience.

Speculations about ultimate "why" questions are enjoyable at the intellectual level.[6] But they also contain more than a whiff of the absurd, trying to peek behind the curtains that hide the origin of creation only to find an endless set of further curtains. I will happily go to my grave knowing that in this universe, IIT characterizes the relationship between experiences and their physical substrate.

Let me now get to the theory. I will be providing a 50,000-foot overview of it, enough, I hope, for you to get a taste of its principles and mode of operation. In no way should these pages be construed as a rigorous, thorough, or exhaustive overview of the theory.[7]

8 Of Wholes

You have arrived at the heart of the book. Come and dive with me into the deep end of the pool.

According to integrated information theory (IIT), consciousness is determined by the causal properties of any physical system acting upon itself. That is, consciousness is a fundamental property of any mechanism that has cause-effect power upon itself. Intrinsic causal power is the extent to which the current state of, say, an electronic circuit or a neural network, causally constrains its past and future states. The more the system's elements constrain each other, the more causal power. This causal analysis can be done for any system with an appropriate causal model, that is, a description of the sort "this widget over here influences the state of that gadget over there in a specific manner," such as a wiring diagram.[1] The theory takes the five phenomenological axioms of experience that I introduced in the first chapter—any experience exists for itself, is structured, is the specific way it is, is one, and is definite—and formulates for each one an associated causal postulate, a requirement that any conscious system has to obey. The theory reveals or unfolds the intrinsic causal powers of any system that obeys all five postulates. These causal powers can be represented as a constellation of points (distinctions) linked by lines (relations). According to IIT, these causal powers are identical to conscious experience, with every aspect of any possible conscious experience mapping one-to-one onto aspects of this causal structure. Everything is accounted for on either side.

All these causal powers, including their degree of intrinsic existence, can be evaluated in a systematic fashion, that is, algorithmically. Figure 8.1 illustrates this structure by way of a visual metaphor, a byzantine tangled spider web.

Let us look at each of the five postulates in turn to see what IIT has to say about them.

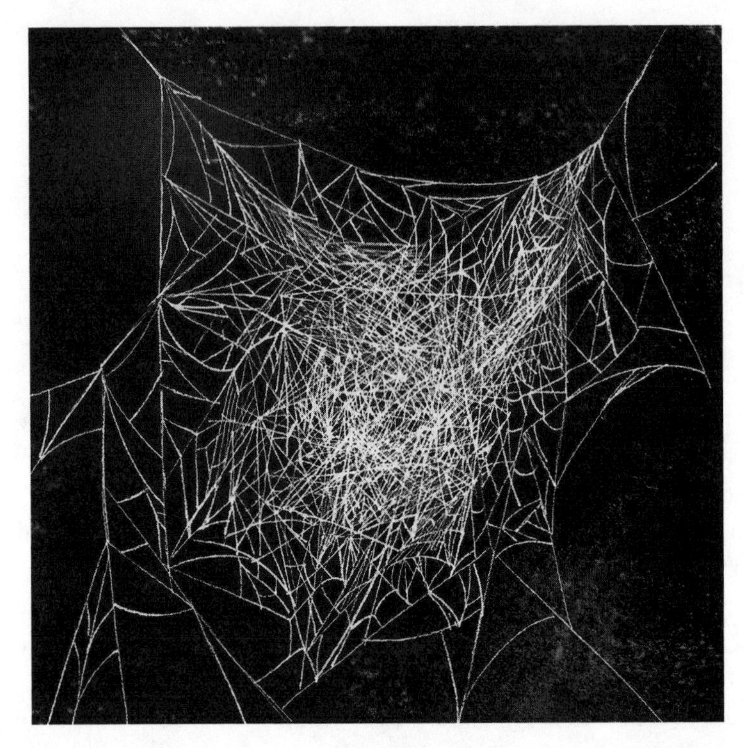

Figure 8.1
A web of causal relations: For any physical system that obeys all five phenomenological axioms, integrated information theory unfolds an associated intrinsic cause-effect structure illustrated using a labyrinthine spider web. The central identify of IIT claims that what it feels like to be this system in this state is identical to the set of causal relationships that compose this structure.

Intrinsic Existence

The starting point of IIT is the Augustinian-Cartesian assertion that consciousness exists intrinsically, for itself, without an observer.[2] The associated *intrinsic existence* postulate contends that to exist intrinsically, any set of physical elements must specify a set of "differences that make a difference" to the set itself.

Switching from an externalist to an internalist perspective sounds straightforward but has profound consequences that inform the entire approach. Most importantly, it removes the point of view of the outside observer, such as a neurophysiologist viewing your brain activity in a scanner and noticing that your fusiform face area is turning on. For consciousness, there is no such observer. Everything must be specified in terms of differences that make a difference to the system itself.

For you to *see* a face, the visual stimulus has to trigger a change that makes a difference to the neuronal substrate constituting your experience. Otherwise, it will be noticed as much as gastric juice trickling into your digestive tract—not at all.

To understand how IIT defines "a difference that makes a difference," I refer to one of the foundational documents of classical philosophy, Plato's *Sophist.* It is a lengthy dialogue between a young mathematician and a stranger from the Greek settlement of Elea in southern Italy, the hometown of Parmenides. At some point, the Eleatic Stranger observes:

> My notion would be, that anything which possesses any sort of power to affect another, or to be affected by another, if only for a single moment, however trifling the cause and however slight the effect, has real existence; and I hold that the definition of being is simply power.[3]

For something to exist from the point of view of the world, extrinsically, it must be able to influence things and things must be able to influence it. That is what it means to have causal power. When something can't make a difference to anything in the world or be influenced by anything in the world, it is causally impotent. Whether or not it exists makes no difference to anything.

This is a widespread but rarely acknowledged principle. Consider aether, a hypothetical space-filling substance or field pervading the entire universe. Luminiferous or light-bearing aether (or ether) was introduced into classical physics in the nineteenth century to explain how light waves propagate through empty space, that is, without a substrate. This required an invisible material that wouldn't interact with ordinary physical objects. When more and more experiments concluded that, whatever the aether might be, it had no effects whatsoever, it finally fell under Occam's razor and was quietly retired. As aether has no causal power, it plays no role in modern physics; and thus it does not exist.

Per the Eleatic Stranger, to exist for others is to have causal power over them. IIT asserts that for a system to exist for itself, it must have causal power over itself. That is, its current state must be influenced by its past and it must be able to influence its future. As the Eleatic Stranger surmised, the more intrinsic causal power something has, the more it exists for itself.

Leibniz wrote about the monad being laden with its past and pregnant with its future. That is intrinsic causal power.

Causal power is not some airy-fairy ethereal notion but can be precisely evaluated for any physical system, such as binary gates implementing Boolean logic or all-or-none neurons in a neural circuit. Everything outside the system, say, other circuits that are connected to it, is considered background (such as the brainstem, as explained in chapter 6) and kept fixed. For if the background conditions change, the causal power of the system might change as well. The question now is "How much is the current state of the system constrained by its past and how much does it constrain its future?"

Let's work through an absurdly simple example by computing the integrated information associated with the circuit of figure 8.2—three logic gates wired up as shown (with the arrows indicating the direction of the causal influence). The OR gate P is ON, equivalent to a logical 1 while the COPY gate Q and the exclusive OR (or XOR) gate R are both OFF, equivalent to a logical 0. The circuit diagram, together with the current state of the system (PQR) = (100), completely specifies the past and future of the circuit.

Given the logical functions carried out by these gates, the circuit will switch from its present state (100) to (001). Updated on a regular clock cycle, this is how all basic computer circuits operate.[4]

Composition

According to the composition postulate, any experience is structured. This structure must be reflected in the mechanisms that compose the system specifying the experience.

To compute the integrated information Φ (the Greek capital letter phi, pronounced *fy*) associated with the triadic circuit of figure 8.2, we have to consider all possible mechanisms—for each one we ask whether it specifies a difference that makes a difference *within* the candidate system. This includes the three elementary gates (P), (Q), and (R) by themselves, the three possible pairs of gates, (PQ), (PR), and (QR), and the three-element circuit (PQR). That is, seven mechanisms have to be evaluated for their integrated information using a unique metric that captures the difference they make from the intrinsic perspective of the system.

In a system that fulfills all postulates, each one of these mechanisms, assuming that they have causal power within the system, that is, non-zero

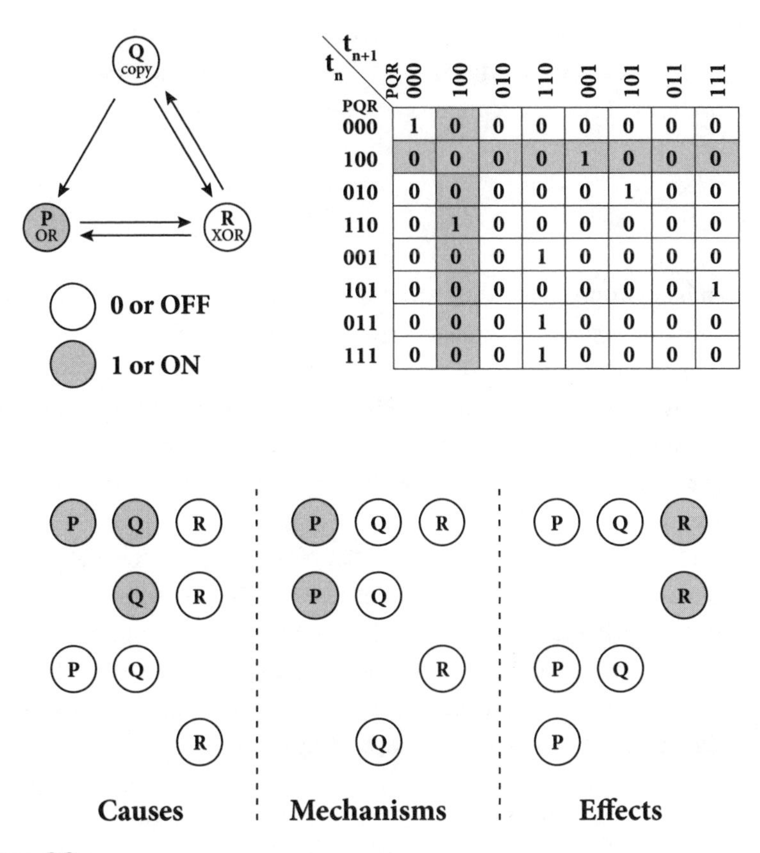

t_n \ t_{n+1} PQR	000	100	010	110	001	101	011	111
000	1	0	0	0	0	0	0	0
100	0	0	0	0	1	0	0	0
010	0	0	0	0	0	1	0	0
110	0	1	0	0	0	0	0	0
001	0	0	0	1	0	0	0	0
101	0	0	0	0	0	0	0	1
011	0	0	0	1	0	0	0	0
111	0	0	0	1	0	0	0	0

Figure 8.2
Causal mechanisms: A circuit of three logic gates (PQR) in the state (100). The table corresponds to the transition probability matrix of this circuit (fully deterministic for ease of exposition). Four mechanisms link four maximally irreducible causes to four maximally irreducible effects that maximally constrain the current state of the circuit. These correspond to four irreducible distinctions. The irreducibility of the entire circuit is quantified by its integrated information Φ.

integrated information, constitute a particular phenomenological *distinction* within an experience. Distinctions are the building blocks of any one experience, bound together by countless *relations* that occur when distinctions overlap or share units. My experience of *Mona Lisa*'s mysterious smile is one higher-order distinction within countless relations that make up the larger visual experience of looking at da Vinci's famous painting; seeing her face, her lips, and so on are other distinctions.

When considering the causal power of the entire circuit, the *composition* postulate requires evaluating all single elements (first-order mechanisms), combinations of two interconnected elements (second-order mechanisms), and the full system consisting of all three elements.

Information

The *information* postulate states that a mechanism contributes to experience only if it specifies "differences that make a difference" within the system itself. A mechanism in its current state specifies information to the extent that it picks out its cause and its effect within the system. A system in its current state generates information to the extent that it specifies the state of a system that can be its possible cause in the past and its effect in the future.

Consider the OR gate P in figure 8.2 that receives inputs from gates Q and R. Since each one of these can be either OFF or ON, there are four possible input states. It is in the nature of an OR gate to be ON if either one of its inputs is ON. Thus, given that P is ON, one of three possible cause states is specified. If, instead, P is OFF, a single cause is specified, namely both inputs being OFF. In another imagined scenario, if the gate P is broken and intrinsic noise makes it equally likely for its output to be OFF or ON, nothing reliably can be inferred about its possible causes. Thus, P's cause-information is maximal when OFF, less when ON and zero when the gate acts randomly.

Similar considerations apply to effect-information in the future. If the present state leads with high probability to one or a few states, the circuit's selectivity and therefore its effect-information is high. If the present only weakly influences the future, say because of high level of ambient noise, or a dependence on other elements within the system, the effect-information is lower.

The cause-effect information is defined as the smaller (minimum) of the cause-information *and* the effect-information. If either one is zero, the cause-effect information is likewise zero. That is, the mechanism's past must be able to determine its present, which, in turn must be able to determine its future. The more the past and the future are specified by the present state, the higher the mechanism's cause-effect power.

Note that this usage of "information" is very different from its customary meaning in engineering and science introduced by Claude Shannon. Shannon information, which is always assessed from the external perspective of an observer, quantifies how accurately signals transmitted over some noisy communication channel, such as a radio link or an optical cable, can be decoded. Data that distinguishes between two possibilities, OFF and ON, carries 1 bit of information. What that information is, though—the result of a critical blood test or the least significant bit in a pixel in the corner of a holiday photo—completely depends on the context. The meaning of Shannon information is in the eye of the beholder, not in the signals themselves. Shannon information is observational and extrinsic.

Information in the sense of integrated information theory reflects a much older Aristotelian usage, derived from the Latin *in-formare,* "to give form or shape to." Integrated information gives rise to the cause-effect structure, a form. Integrated information is causal, intrinsic, and qualitative: it is assessed from the inner perspective of a system, based on how its mechanisms and its present state shape its own past and future. How the system constrains its past and future states determines whether the experience feels like azure blue or the smell of wet dog.

Integration

Any conscious experience is unified, holistic. The associated *integration* postulate requires that the cause-effect structure specified by the system be unified or irreducible. Integrated information Φ quantifies to what extent the form that is generated by the whole is different from that generated by its parts. That is the meaning of irreducible—the system can't be reduced to independent, noninteracting components without losing something essential. Irreducibility is evaluated by considering all possible partitions of the circuit, all the different ways the circuit can be decomposed into nonoverlapping mechanisms (which can be of very unequal size).[5] The actual number Φ quantifies the extent to which the cause-effect structure changes if it is cut or reduced along its minimum information partition (the cut that makes the least difference). Φ is a pure number that is either zero or positive.[6]

For instance, if the two connections between P and R and between P and Q are severed, the cause-effect structure of the two separate mechanisms is quite different from the one associated with the entire circuit, as interdependencies between the two mechanisms are no longer captured.

If partitioning some entity makes no difference to its cause-effect structure, then it is fully reducible to those parts, without losing anything. In a very real sense it does not exist as a system but only as disjoint parts. Its Φ is 0. Think of citizens trying to organize as a group to fight the construction of a planned freeway through their neighborhood. If they never meet, never interact, and don't coordinate their actions, then their group doesn't exist from the point of view of its external causal effect in local politics. For an example of intrinsic causal power, consider your and my consciousness. You can have a painful experience and so can I, perhaps even at the same time. But there is no sense in which there is an experience associated with the pair of us. Our joint experience is fully reducible to that of you and me.

Exclusion

This type of causal analysis shows that the system (PQR) is irreducible, that is, it can't be reduced to two or more components without losing something in the process.

We repeat this computation for all possible combinations of gates as candidate circuits—not just the triad (PQR), but also the pairs (PQ), (PR) and (QR) and the elementary gates themselves. Each of these seven "circuits" has its own cause-effect structure, with its own Φ value. The phenomenological axiom that any experience is definite is associated with the *exclusion postulate* that stipulates that only the circuit that is maximally irreducible exists for itself, rather than any of its supersets or subsets. All overlapping circuits with smaller values of Φ are excluded.

The fact that only the maximum exists from an intrinsic point of view appears strange to many.[7] However, there are any number of extremal principles in physics. Take the principle of least action, a key theme in relativity theory, thermodynamics, mechanics, and fluid mechanics. It stipulates that of all the ways that a particular physical system can evolve, only one actually occurs, the one that is an extremum. For instance, the shape of a sagging bicycle chain or a metal ring chain suspended at both ends is the one that minimizes its potential energy.[8]

The maximum of Φ is a global maximum over the substrate considered. That is, there will not be any superset or subset of this circuit with more integrated information. There will be, of course, plenty of nonoverlapping systems, such as other brains, that will have higher Φ values.

The exclusion postulate provides a sufficient reason for why the content of an experience is what it is—neither less nor more. With respect to causation, this has the consequence that the winning cause-effect structure excludes any alternative cause-effect structures specified over overlapping elements, as otherwise there would be causal overdetermination. This exclusion of causes and effects is another form of Occam's razor: "Causes must not be multiplied beyond necessity."

Only the triad (PQR) survives this vetting process, as it has the largest Φ value, called Φ^{max} (phi-max), given the background conditions. If these change, Φ^{max} is likely to change as well. From an intrinsic perspective, (PQR) is irreducible, its borders defined by the set of elements that yield a maximum of Φ.

The theory calls this set of elements the *main complex* or the *physical substrate of consciousness*. I give it the more poetic name, the *Whole* (with a capital W). The Whole is the most irreducible part of any system, the one that makes the most difference to itself.[9]

Per IIT, only this Whole has an experience. All others, such as the smaller circuit (PQ), do not exist for themselves, as they are not a maximum of Φ; they have lesser intrinsic causal power.

The Central Identity of Integrated Information Theory

The elements of a Whole in some state, alone and in combination, compose first- and higher-order mechanisms that specify distinctions. All these mechanisms, bound together by relations, form a structure, defined as the maximally irreducible cause-effect structure. Given the set of all possible irreducible first- and higher-order mechanisms and their overlap, the complexity of this structure for any realistic circuit will be mind boggling—the spider web of figure 8.1 gives a pale impression of this.

The theory has a startlingly precise answer to the question of what consciousness is, in its *central identity*:

> Any experience is identical to the maximally irreducible cause-effect structure associated with the system in that state.

The experience is identical to this structure—not to its physical substrate, its Whole, just as my experience of feeling kind of blue is not identical to the gray goo in my head that is the physical substrate of this experience.

The completely unfolded maximally irreducible cause-effect structure is an entity with particular causal properties. It corresponds to a "constellation" of points bound by relations, or to a *form*.[10] It is not an abstract mathematical object nor a set of numbers. It is physical. Indeed, it is the most real thing there is. I called this form a *crystal* in my earlier book:

> The crystal is the system viewed from within. It is the voice in the head, the light inside the skull. It is everything you will ever know of the world. It is your only reality. It is the quiddity of experience. The dream of the lotus eater, the mindfulness of the meditating monk, and the agony of the cancer patient feel the way they do because of the shape of the distinct crystals in a space of a trillion dimensions—truly a beatific vision.[11]

For brains, the Whole is the neural correlate of consciousness at the relevant level of granularity whose state can be observed and manipulated (more on this in the next two chapters). The Whole has definite membership, with some neurons belonging while others, perhaps intimately connected to the former, do not. For the toy network of figure 8.2, the Whole is the triadic circuit (PQR). External ports that feed into this circuit or that read out its state are not part of its physical substrate of consciousness. Its experience is given by the constellation of four distinctions bound by their causal relations within the maximally irreducible cause-effect structure (fig. 8.2). The larger the system's irreducibility Φ^{max}, the more it exists for itself, the more it is conscious. Φ^{max} has no obvious upper bound.[12]

The central identity of IIT, a metaphysical statement, makes a strong ontological claim. Not that Φ^{max} merely *correlates* with experience. Nor the stronger claim that a maximally irreducible cause-effect structure is a necessary and sufficient condition for any one experience. Rather, IIT asserts that any experience is identical to the irreducible, causal interaction of the interdependent physical mechanism that make up the Whole. It is an identity relationship—every facet of any experience maps completely onto the associated maximally irreducible cause-effect structure with nothing left over on either side.

The claim that it feels like something to be (PQR) is a far-out extrapolation of the theory, as counterintuitive as the claim that humans, worms, and giant Sequoia trees are all evolutionarily related sounded to most people in the Victorian age. This claim will have to be cashed in. It will be essential to demonstrate, using the theory's tools and concepts, how the way any one experience feels is explained by different aspects of the associated cause-effect structure. In this context, consider the seemingly "simple" experience of seeing the text "Wake up Neo" on a computer monitor. In fact, there is absolutely nothing simple about this experience, as its phenomenological content is incomprehensible large, composed of many distinctions and relations. These make this experience what it is and different from any other.

These distinctions include the lower-order ones specifying the individual pixels, the way they form the oriented edges making up the individual letters at a particular location, as well as the higher-order invariant distinctions that make up the letter "W" (a specific arrangement of lines at particular angles in any of a large number of possible locations) and the individual words. But there is more, so much more, for the very experience of seeing something, anything, spatial is highly meaningful—containing a multitude of distinctions (points) extended in space, over neighborhoods, near and far, to the left and the right, up and down, and so on. These distinctions are bound together within the same experience in a complex pattern of relations—the letters are located somewhere, in a particular font, capitalized or not, in a specific relationship to words and a name that is, itself, a high-level distinction. According to the theory, such a dynamic "binding" of perceptual attributes[13] occurs if distinctions share overlapping set of mechanisms jointly constraining their past and future states.

Adopting the intrinsic point of view means rejecting the external "God's-eye" perspective that sees something as spatially extended. It means specifying the causal powers that constitute the way space—empty or not—looks and feels.

How This Works in Practice

To make this concrete, let us consider the set of neurons making up the posterior neocortical hot zone. Let us assume a specific state in which some of these neurons are firing over some time window (say 10 milliseconds), while most are silent. The rest of the brain is treated as a fixed background; that is, the firing of the neurons in the front of cortex, in the cerebellum, brainstem and so on are held constant at whatever values they have.

The challenge is to discover those neurons within the hot zone that are the substrate for the maximally irreducible cause-effect structure, the Whole, and to compute its Φ^{max}.

We start with a causal model of the network, a wiring diagram that specifies how individual neurons within the posterior hot zone are connected, with what weights and firing thresholds. Of course, given today's inadequate state of knowledge, much here has to be surmised (for instance, a grid-like connectivity compared to more random access connectivity for the front of cortex). With such a diagram in place, we compute the cause-effect information, starting with first-order mechanisms, individual neurons.

Any neuron can have cause-power, that is, can constrain the states of its inputs. This is evaluated, for all possible subsets of presynaptic neurons, by specifying a cause-state and an effect-state (or, more simply, specifying a cause and an effect). The integrated cause-information is the weighted difference between the partitioned and unpartitioned causes (selected as the partition yielding the *smallest* distance); it measures what difference the cause makes to the mechanism in question, assessing the reducibility of that neuron's cause-power.

The core-cause of a neuron is that set of inputs with the largest cause-power. A similar procedure determines the core-effect of the neuron among its synaptic outputs. A neuron that has both a non-zero core-cause and core-effect specifies a distinction (e.g., fig. 8.2).

We do this for all possible mechanisms, that is, for all individual neurons, for all combinations of two neurons (second-order mechanism), of three neurons (third-order mechanism), and so on, up to the entirety of the network. Yes, that is a large number. (However, we need not consider the vast subset of these mechanisms that don't directly interact.) This accounts for all possible distinctions of the network (the circuit in figure 8.2 has four core-causes and core-effects, with four distinctions).

The irreducibility of a network is a measure of the network's overall integration. This is assessed similarly to the way in which we assessed a single neuron's integration: partition the network and measure the extent to which it can be recovered from the resulting parts. Φ is a scalar measure of the irreducibility of the system over all possible subsets of the network.

Many networks within the hot zone will have positive Φ values. Per exclusion, though, only the circuit with the maximum of Φ over the network is a Whole, existing for itself.

Practically, a system with intrinsic, irreducible cause-effect power must be strongly connected. But a fully connected network, in which every unit is wired to every other one, is not the best way to achieve highly integrated information.

For those conversant with the open-source programming language Python, a publicly available software package that computes the Whole and Φ^{max} for small networks can be freely downloaded.[14]

If you've made it through this chapter, congratulations, as the covered material stretches anyone's mind. In the next two chapters, I will discuss clinical implications of the theory, such as a consciousness meter, and some quite counterintuitive predictions. This provides further insight into how the theory works.

9 Tools to Measure Consciousness

I have no doubt that when you tell me "I vividly recall where I was when I first saw the burning and crumbling Twin Towers on TV," you are conscious. Language is the gold standard of inferring consciousness in others. However, this doesn't work for those incapacitated and unable to speak. How can an outsider know whether or not they are conscious? This is a daily dilemma in the clinic.

Consider patients put to sleep during surgery to implant a stent, remove a cancerous growth, or replace a worn-out hip. Anesthesia eliminates pain, prevents the formation of traumatic memories, keeps patients immobile, and stabilizes their autonomic nervous system. Patients are put under in the expectation that they won't wake up during surgery. Unfortunately, this goal is not always met. Intraoperative recall, or "awareness under anesthesia," occurs in a small fraction of operations, estimated to be in the one to a few per 1,000 range. As patients are paralyzed to facilitate intubation and prevent gross muscle movement, they can't signal their distress. Given that 50,000 or more Americans undergo general anesthesia each day, this tiny fraction translates into hundreds of awakenings under anesthesia.

Another group whose state of consciousness is uncertain are patients with immature, deformed, degenerated, or injured brains. Young or old, they are mute and don't respond to verbal requests to move their eyes or limbs. Establishing that they experience life is a grave challenge to the clinical arts.

It is difficult to be confident about the facts of consciousness in preterm babies living in an incubator, severely malformed infants such as those without a cortex, skull, and scalp (*anencephaly*[1]), or those born from mothers infected with the Zika virus (*microcephaly*). Indeed, until the closing years of the twentieth century, preemies were often operated on without anesthesia, to minimize acute and long-term risk to their very fragile bodies, and because their brains were thought to be too immature to experience pain.[2]

At the other end of life are elders with severe dementia. The final stage of Alzheimer's and other neurodegenerative diseases is marked by extreme apathy and exhaustion. Individuals cease speaking, gesturing, and even swallowing. Has their conscious mind permanently left its abode, a shrunken brain full of neurofibrillary tangles and amyloid plaques?[3]

What would be helpful in all these cases is an instrument like the tricorder from *Star Trek* that signals the presence of experience, a consciousness meter. David Chalmers introduced this concept in a lively talk during which he pointed a hair-dryer at various people to pretend to unmask zombies in the audience.[4] However imperfect, such a tool would drive progress. Let me discuss a way to detect the echoes of a conscious mind by probing the brain. But before I do, I will introduce the patients who would most benefit from such an instrument.

Stranded Minds in Damaged Brains?

The last half-century has witnessed a remarkable revolution that plucked thousands of victims of acute and massive brain injuries from the shores of death and returned them to the lands of the living. Before the advent of advanced medical care, surgical and pharmacological intervention, emergency helicopters, 911, and so on, these patients would have quickly succumbed to their injuries.[5]

However, the dialectic of progress has a dark underbelly: patients who remain bedridden and disabled, unable to articulate their mental state, with ambiguous and fluctuating signs of conscious perception of self and their environment, in a diagnostic limbo as regards to their degree of consciousness. As this state can last years, it imposes a terrible burden on the loved ones who care for them, unrelieved by any emotional closure that death and the subsequent grieving process can bring.

Patients with *disorders of consciousness* are a diverse group.[6] It helps to think about them using two criteria. One is the extent to which they respond to relevant external cues in purposeful ways, such as moving their eyes, nodding, and so on. The other is the extent of any remaining cognitive capabilities. Can they still recall, decide, think, or imagine? Each patient lives in a

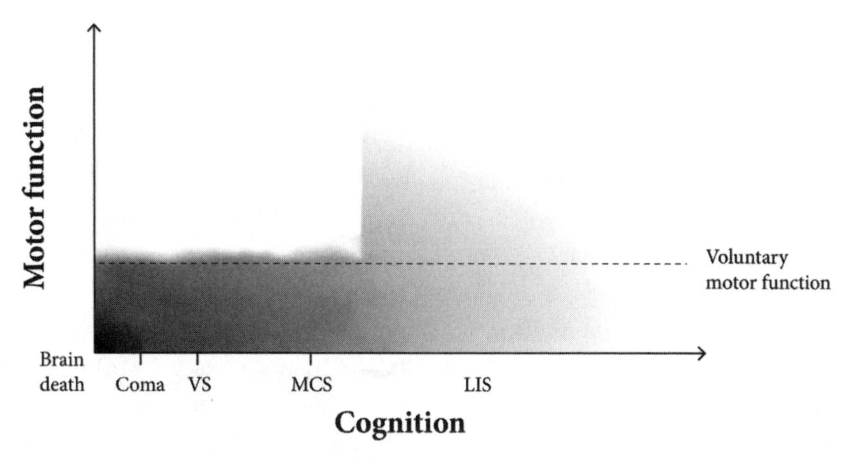

Figure 9.1

Broken brains and consciousness in the plains of misery: Classifying brain-injured patients with disorders of consciousness as a function of their residual cognitive capabilities (progressing horizontally, from none on the left to intact processing in locked-in patients (LIS) on the right) and residual motor function (changing along the vertical axis from none at the bottom to reflexes, to goal-directed movements at the top). A critical determination is whether patients can respond voluntarily (dashed line) with their eyes, limbs or speech. The darker the shading, the more likely consciousness is absent. (Based on the work of the neurologist Niko Schiff [e.g., Schiff, 2013].)

particular spot in these plains of misery (fig. 9.1), in which the horizonal extent represents their cognitive abilities and the vertical axis the extent of their motor actions.

Their abilities to act are judged by closely observing them—do they startle in response to loud tones? Do they withdraw the hand or leg if pinched? Do their eyes track a bright light? Do they initiate any nonreflex behaviors? Can they signal with hands, eyes, or head in response to a command? Are they capable of any meaningful utterances?

Inferring the extent to which patients retain residual cognitive skills is trickier when they can't speak. A typical clinical assessment includes astute bedside observation—some locked-in patients are fully conscious yet can only signal by blinking their eyes—or even placing the patient into a brain scanner and asking them to imagine playing tennis or visiting, in their mind, each room in their home. These two tasks selectively increase blood flow to one of two distinct cortical regions. Patients who can willfully regulate their brain activity in this manner are taken to be conscious.[7]

Healthy, regular folks who walk about and speak cluster in the upper right-hand corner of figure 9.1, with high cognitive and motor capabilities. The extreme opposite is at the lower left-hand corner, marked by a black hole, *brain death*. This is defined as the complete and irreversible loss of central nervous system function, or total brain failure. No behavior, whether automatic, reflex-like or voluntary, remains.[8] No more experience, either. Once inside this black hole, there is no escape.

In the United States, as in most advanced countries, whole brain death legally is death tout court. That is, when your brain is dead, you are dead. This diagnosis can be difficult for family and friends to accept at the bedside, as the patient, now technically a corpse, will often breathe, owing to mechanical venting, and have a beating heart. At this point, the body is an organ donor.

There are, however, credible reports of "life" after death. Though the body is legally speaking a corpse (but on life support), the skin retains a healthy complexion, fingernails and hair grow, the body is warm, and menstruation may occur. Such cases demonstrate how medicine, science, and philosophy continue to struggle to coherently define that transition from life to non-life that we eventually all undergo.[9]

In terms of loss of function, *coma* is a close cousin to brain death. The comatose patient is alive yet immobile, with eyes closed even when presented with the most vigorous stimulation; only a few brainstem reflexes remain. Unless pharmacologically maintained, coma is typically short-lived, ending either in death or in partial or full recovery.

Next is the *vegetative state* (VS), which I wrote about in chapter 2. VS patients, in contrast to comatose ones, have irregular cycles of eye opening and closure. They can swallow, yawn, may move their eyes or head but not in an intentional manner. No willed actions are left—only activity that controls basic processes such as breathing, sleep–wake transitions, heart rate, eye movements, and pupillary responses. Bedside communication—"If you hear me, squeeze my hand or move your eyes"—meets with failure. With proper nursing care to avoid bedsores and infections, VS patients can live for years.

Their inability to communicate in any way, shape, or form is compatible with the notion that VS patients do not experience anything. Yet recall the mantra "Absence of evidence is not evidence of absence." There is a diagnostic gray zone into which VS patients fall as far as the crucial question of whether their injured brains experience pain, distress, anxiety, isolation, quiet resignation, a full-blown stream of thought—or nothing at all. Studies suggest that perhaps 20 percent of VS patients are conscious and have therefore been misdiagnosed.

To family and friends, who may care for their loved one for years, knowing whether anybody is home can make a dramatic difference. Think of an astronaut untethered from his safety line, adrift in space, listening to mission control's increasingly desperate attempts to contact him ("Can you hear me, Major Tom?"). But his damaged radio doesn't relay his voice. He is lost to the world. This is the desperate situation of some patients whose damaged brain won't let them communicate, an involuntary and extreme form of solitary confinement.

The situation is less ambiguous for *minimally conscious state* (MCS) patients. Unable to speak, they can signal but often only in a sparse, nominal, and erratic fashion, smiling or crying in appropriate emotional situations, vocalizing or gesturing on occasion, tracking salient objects with their eyes, and so on. While the evidence suggests that these patients likely experience something, it is an open question how fleeting and diminutive their experiences are.

Locked-in syndrome (LIS) patients are afflicted by bilateral brainstem lesions that prevent most or all voluntary movement while preserving consciousness (lower right-hand side of figure 9.1). Often their sole remaining link to the world are vertical eye or minute facial movements. The French writer Jean-Dominique Bauby dictated a short book, *The Diving Bell and the Butterfly*, by blinking with his left eyelid, while the British cosmologist Stephen Hawking was a genius entombed inside a body crippled by progressive motor neuron disease.[10]

In completely locked-in patients, such as those in end-stage amyotrophic lateral sclerosis (Lou Gehrig's disease), all links to the outside world are down. In these tragic cases of complete paralysis, knowing whether any mental life remains is extremely challenging.[11]

E Pluribus Unum

Conscious experiences, thoughts, and memories arise and disappear within fractions of a second. Measuring their lithe neuronal footprints requires the use of instruments capable of capturing this dynamic. For the clinical neurologist, this is the electroencephalogram (EEG; fig. 5.1).

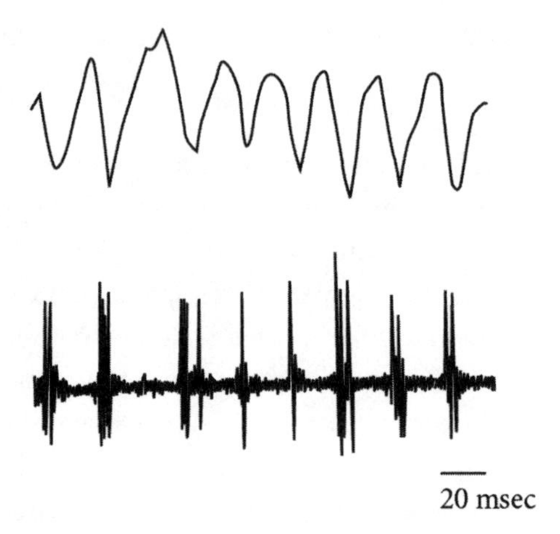

Figure 9.2
Gamma oscillations, a hallmark of consciousness? Brain signals are picked up by a fine microelectrode inserted into the visual cortex of a cat looking at a moving bar. The local field potential (the summed electrical activity of cells around the electrode in the top trace) and spiking activity from a handful of nearby neurons (bottom trace) has a prominent periodicity in the 20–30 msec range. Crick and I argued that this was a neural correlate of consciousness. (Modified from Gray & Singer, 1989.)

From the late 1940s onward, an *activated EEG* was the surest sign of a conscious subject. This is characterized by low-voltage, rapidly fluctuating waves that are desynchronized, that is, not in lock-step, across the skull. As the EEG shifts to lower frequencies, consciousness is less likely to be present, and the brain is asleep, sedated, or damaged.[12] Yet there are enough exceptions that this rule cannot serve as a general indicator of the absence or presence of consciousness in any one individual. Thus, basic scientists and clinicians alike have cast about for more reliable measures.

Crick's and my quest for the neuronal footprints of consciousness galvanized around the (re)discovery in 1989 of synchronized discharges in the visual cortex of cats and monkeys looking at moving bars and other stimuli (fig. 9.2). Neurons fired action potentials not haphazardly but in a periodic manner, with a pronounced tendency for spikes to be spaced 20–30 milliseconds apart. These waves in the recorded electrical potential are known as gamma oscillations (30–70 cycles per second or Hz, roughly centered around 40 Hz, equivalent to a 25 millisecond periodicity). Even more remarkably, nearby neurons that signaled the same object fired roughly simultaneously. This led us to propose that these 40 Hz oscillations are an NCC.[13]

This simple idea captured the attention of scientists, powering the modern quest for the NCC. Yet after more than a quarter of a century of empirical investigations of 40 Hz oscillations in people and animals in hundreds of experiments, the verdict is that they are not a true NCC! What emerged is a more nuanced view of the relationship between gamma oscillations and consciousness. Periodic firing of neurons in this range or gamma band activity in the EEG (upper left trace in fig. 9.1) is closely linked to selective attention, with synchronized gamma oscillations between two regions strengthening the effective connectivity within the two underlying neuronal coalitions. Just as we can pay attention to something that we don't see (chapter 4), stimuli can trigger 40 Hz oscillations without evoking any conscious experience.[14]

Yet gamma oscillations continue to surface in connection with consciousness. A commercial system preferred by many anesthesiologists for its ease of use is the bispectral index (BIS) monitor. It evaluates the EEG for a preponderance of high-frequency components relative to low-frequency ones (the exact details are proprietary). In practice, this doesn't help to reduce how often patients wake up during anesthesia with subsequent recall.[15] Further, the BIS does not work in a consistent manner across the wide range of patients put under, from neonates to the elderly, nor across the diversity of anesthetic agents in existence, nor is it relevant to the neurological patients described above, whose EEG patterns can be abnormal. Other measures, such as the P3b, are no better at marking consciousness.[16]

All these indices are, by and large, based on the analysis of the time-course of a single electrical signal. Yet modern EEG systems simultaneously record the voltage at sixty or more locations on the scalp (called "channels" in the lingo). That is, the EEG has a spatiotemporal structure. Except when the EEG is flat, as during deep anesthesia, or when waves that goose-step across the entire cortex appear during coma or status epilepticus, the EEG has a complex appearance, revealing its dual, integrated, and differentiated character. Integrated in the sense that signals at different locations are not fully independent of each other; differentiated in the sense that they each are highly structured in time, with their own individuality.

Integrated information theory implies that the neural correlates of consciousness should mirror the unitary aspect of each experience as well as reflecting the highly diverse nature of any one experience. Applying these principles to the EEG's spatiotemporal structure yields a tool that reliably detects the presence of consciousness in everyone.

The spirit of this technique, dubbed "zap-and-zip" for reasons that will become apparent, is the same as the official motto of the United States, *e pluribus unum*: out of many, one. It was devised by Giulio Tononi, the creator of IIT, and the neurophysiologist Marcello Massimini, now at the University of Milan in Italy.[17] An enclosed coil of wire held against the scalp sends a single, high-field pulse of magnetic energy into the neural tissue beneath the skull, inducing a brief electrical current in nearby cortical neurons and their axons via electromagnetic induction.[18] This, in turn, briefly engages their synaptically connected partner cells in a cascade that reverberates inside the brain before it dies out within a fraction of a second. The resultant electrical activity is monitored by a high-density EEG cap. Zapping the brain with many such pulses, averaging the EEG, and unfolding it over time yields a movie. Figure 9.3 illustrates the EEG record following one magnetic zap.

The spatiotemporal activity in response to the magnetic pulse is a highly complex pattern, waxing and waning as the waves of excitation ricochet across the tissue, spreading from the trigger site. Think of cortex as a large

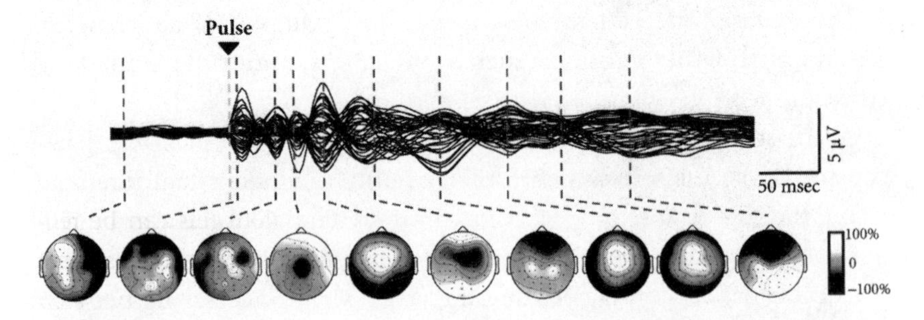

Figure 9.3
Zapping the brain: Provoking the brain with a magnetic pulse generates a short-lived wave of neural agitation that perturbs the ongoing cortical activity measured underneath 60 EEG channels in an awake volunteer (bottom). The averaged electrical signals from all channels are displayed at the top. Their spatiotemporal complexity is characterized by a single number, the perturbational complexity index or PCI, to infer consciousness. (Modified from Casali et al., 2013.)

brass bell and the magnetic coil as the clapper. Once struck, a well-cast bell will ring at its characteristic pitch for a considerable time. And so does the awake cortex, buzzing at a variety of frequencies.

In contrast, the brain of someone in a deep sleep acts like a stunted or cracked bell (like the Liberty Bell in Philadelphia). Whereas the initial amplitude of the EEG during sleep is larger than when the subject is awake, its duration is much shorter, and it does not reverberate across cortex to other connected regions. Although the neurons remain active, as evidenced by the strong, local response, integration has broken down. We see little of the spatially differentiated and temporally variegated electrical activity typical for the awake brain. The same is also true of subjects who volunteered to undergo general anesthesia. The magnetic pulses invariably produce a simple response that remains local, indicative of a breakdown in cortico-cortical interactions and a lessening of integration, in agreement with IIT.

The researchers estimate the extent to which this response differs across cortex and across time using a mathematical measure capturing how compressible the response is. The algorithm is borrowed from computer science and is the basis of the popular "zip" compression algorithm for reducing the storage demand of images or movies, which is why the entire procedure is known in the trade as zap-and-zip. Ultimately, each person's EEG response is mapped onto a single number, the perturbational complexity index (PCI). If the brain barely reacts to the magnetic pulse—say, because the EEG is nearly flat (as in deep coma), its compressibility will be high and PCI will be close to zero. Conversely, maximal complexity yields a PCI of one. The larger the PCI, the more diverse the brain's response to the magnetic pulse, the more difficult it is to compress.

The logic of the approach is now straightforward. In a first step, apply zap-and-zip to a benchmark group of people to infer a constant threshold value, PCI*, such that in every case in which consciousness can be reliably established, PCI is above PCI* and in every case in which the subject is unconscious, PCI values are below this threshold. This establishes PCI* as the minimum complexity value supporting consciousness. Then, in a second step, use this threshold to infer whether consciousness is present in patients in the gray zone.

The benchmark group included 102 healthy volunteers and 48 responsive brain-injured patients. The conscious group was made up of subjects who were awake, some who were dreaming during REM sleep, and some who were under ketamine anesthesia. For the latter two, consciousness was assessed afterward—in sleepers, by randomly awakening them during REM sleep and only including their EEGs if they reported any dream experience immediately prior to awakening; and similarly for subjects under ketamine anesthesia, a dissociative agent that disconnects the mind from the external world but does not extinguish consciousness (indeed, at a lower dose, ketamine is abused as a hallucinogenic drug, "vitamin K"). Unconscious conditions included deep sleep (reporting no experiences immediately prior to being awakened) and surgical-level anesthesia using three very different agents (midazolam, xenon, and propofol).

In every one of these 150 subjects, consciousness was inferred with complete accuracy using a PCI* value of 0.31. Everybody, healthy volunteer or brain-injured patient, was correctly classified as conscious or unconscious. This is a remarkable achievement given the great variability in gender, ages, and medical and behavioral conditions. This was also true for nine patients who recently "woke up," emerging from a vegetative state to become minimally conscious (EMCS), and five locked-in state patients. Each one struggled somewhere on the plains of misery, modified from figure 9.1 by plotting complexity of behavior on the vertical axis and their PCI index on the horizonal one (fig. 9.4).

The team then applied zap-and-zip with this threshold value to minimally conscious-state (MCS) and vegetative state (VS) patients (fig. 9.4). For the former, with some signs of nonreflexive behaviors, the method correctly assigned consciousness to 36 out of 38 patients, but failed in two, misdiagnosing them as unconscious. Of the 43 vegetative-state patients unable to communicate, 34 had EEG responses that were less complex than those of any of the benchmark conscious population, as expected. More troubling, however, were nine patients who responded to the magnetic pulse with a reverberatory pattern whose perturbational complexity was as high as in conscious controls. If the theory is correct, it implies that these patients are conscious yet unable to communicate with the world.

Toward a True Consciousness Meter

These are exciting times—the first principled method for detecting the loss and recovery of consciousness due to the breakdown and recuperation of cortex's capacity for information integration. The technique probes

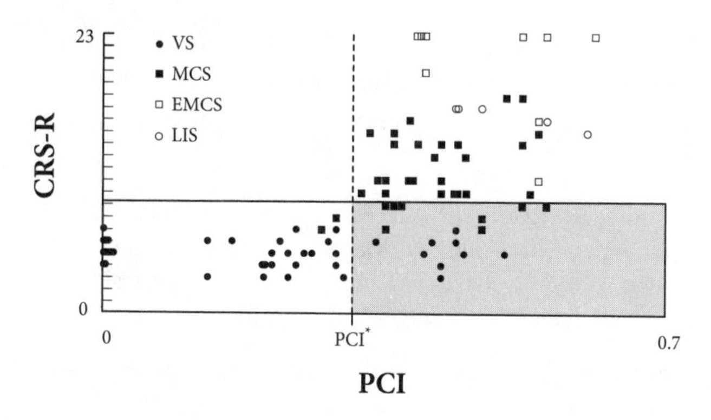

Figure 9.4

Identifying conscious patients using zap-and-zip: The behavior of 95 brain-injured patients as a function of their perturbational complexity index (PCI; fig. 9.1). The dotted line corresponds to a threshold, PCI*, below which consciousness is absent. The vertical axis shows the coma-recovery-scale revised (CRS-R), rating the extent to which patients can respond with their eyes, limbs or speech. Low values indicate reflexes while higher values are associated with more cognitive responses. VS stands for vegetative state, MCS for minimum conscious state, EMCS for emerging minimum conscious state and LIS for locked-in state patients. (Redrawn from Casarotto et al., 2016.)

the intrinsic excitability of cortex, independent of sensory input or motor output, diagnosing patients unable to signal their state in any way. A large-scale collaborative trial in clinics across the United States and Europe seeks to standardize, validate, and improve zap-and-zip so that it can be quickly and reliably administered at the bedside or in the controlled chaos of an emergency room by technicians.

Assigning somebody to the vegetative-state category—unable to initiate any actions that could reveal a mind residing within the damaged brain—is always provisional. Indeed, I just mentioned nine out of 43 patients whom the zap-and-zip method diagnosed as conscious, though they showed no behavioral signs of consciousness as assessed by the clinical gold standard (the coma recovery scale-revised; fig 9.4). This challenges clinicians to devise more sophisticated physiological and behavioral measures to detect the faint telltale signs of a mind using advanced brain–machine interfaces and machine learning.

Zap-and-zip is being extended to catatonic and other dissociative psychiatric patients, the left and right hemispheres of split-brain patients, geriatric patients with late-stage dementia, pediatric patients with smaller and less-developed cortices, and laboratory animals such as mice. I'm so optimistic about this research that I've committed myself to a public wager with the Swedish journalist and science writer Per Snaprud that, by the end of 2028,

> clinicians and neuroscientists will develop well validated brain-activity measurement technologies, establishing with a high degree of certainty whether individual human subjects, such as anesthetized and neurological patients, are, at that moment, conscious.[19]

Zap-and-zip does not measure integrated information. While the PCI index is motivated by IIT, it crudely estimates differentiation and integration.[20] *Pace* IIT, a genuine consciousness meter should measure Φ^{max}. Such a *Φ-meter* or *phi-meter* must causally probe the NCC at the relevant level of spatiotemporal granularity that maximizes integrated information. This can be assessed empirically.[21]

A true phi-meter should reflect the waxing and waning of experience during wakefulness and sleep, how consciousness increases in children and teenagers until it reaches its zenith in mature adults with a highly developed sense of self, with, perhaps, an absolute maximum in long-term meditators, before it begins its inevitable decline with age. Such a device would generalize across species, whether or not they have a cortex or, indeed, any sort of sophisticated nervous system. For now, we are far from such a tool. In the interim, let us celebrate this milestone in the millennia-old mind–body problem.

10 The Über-Mind and Pure Consciousness

An appealing feature of integrated information theory is its fecundity in spawning new ideas such as the primitive but effective consciousness meter I just discussed. Others pertain to the relevant spatiotemporal scale of granularity that maximize integrated information, which is the level at which experience occurs (see note 21 in the last chapter), the neural correlate of pure experience, and the creation of a single conscious mind by merging two brains, one of the theory's more startling predictions. Before I go there, let me start out by discussing the opposite: splitting one brain into two minds.

Splitting a Brain Creates Two Conscious Wholes

Consider the two cortical hemispheres, linked by the two hundred million axons making up the corpus callosum and a few other minor pathways (fig. 10.1). These connections enable both sides of the cortex to smoothly and effortlessly be woven into one mind.

In chapter 3, I mentioned split-brain patients, whose corpus callosum is severed to alleviate seizures.[1] Remarkably, once patients recover from this massive surgical intervention, they are inconspicuous in everyday life. They see, hear, and smell as before; they move about, talk, and interact appropriately with other people; and their IQ remains unchanged. Cutting the brain along its midline does not appear to dramatically affect patients' sense of self or the way they experience the world.

But their innocuous behavior hides a remarkable fact: the speaking mind of the patient almost always has its home in the left cortical hemisphere. Only its experiences and memories are publicly accessible via language. The mind of the nondominant right hemisphere is more difficult to reach; for

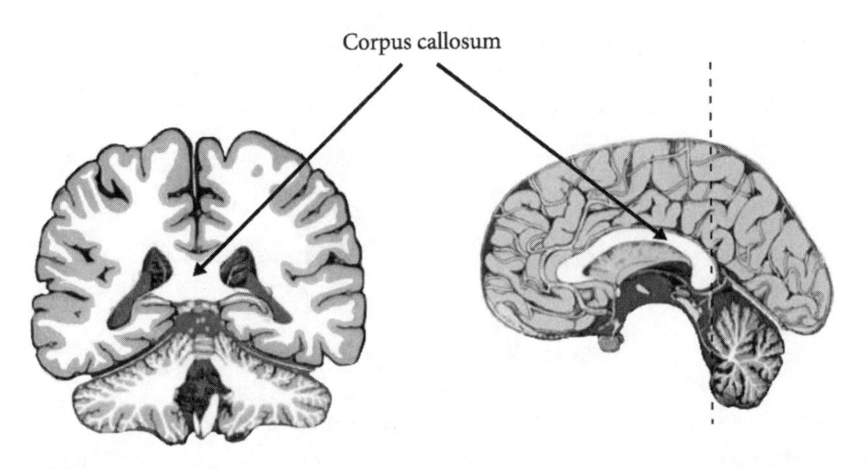

Figure 10.1
A high-bandwidth connection between cortical hemispheres: Two hundred million axons connect the two cerebral hemispheres. (Modified from Kretschmann & Weinrich, 1992.)

instance, by silencing its domineering left partner using anesthesia.[2] Now, the almost mute hemisphere reveals itself by carrying out complex, nonstereotypical behaviors; it can read single words and, in some cases at least, understand syntax and produce simple speech, follow instructions, and sing. As far as outsiders can tell, in split-brain patients there are two distinct streams of consciousness within one skull.[3]

Roger Sperry, who was awarded the 1981 Nobel Prize for his work on these patients, unequivocally states:

> Although some authorities have been reluctant to credit the disconnected minor hemisphere even with being conscious, it is our own interpretation based on a large number and variety of nonverbal tests, that the minor hemisphere is indeed a conscious system in its own right, perceiving, thinking, remembering, reasoning, willing and emoting, all at a characteristically human level, and that both the left and the right hemisphere may be conscious simultaneously in different, even in mutually conflicting, mental experiences that run along in parallel.[4]

How is this interpreted within IIT? Per its central identity thesis, any experience is identical to the brain's maximally irreducible cause-effect structure (fig. 8.1). How much it exists for itself, its irreducibility, is given by the maximum of integrated information, Φ^{max}. The physical structure that determines this experience, its Whole, is the operationally defined content-specific neural correlate of consciousness (the NCC of chapter 5). Its background conditions are all the physiological events that support it—a beating heart, lungs supplying oxygen to the neural tissue, various ascending systems such as noradrenaline and acetylcholine fibers, and so on.

In a normal brain with its two tightly linked cortical hemispheres, the Whole will extend across both hemispheres, with its associated integrated information Φ^{max}_{both} (fig. 10.2). This Whole has sharp boundaries, with some neurons in and the rest out, reflecting the definite nature of every experience. A sheer uncountable number of overlapping circuits have less integrated information. In particular, there will be the Φ^{max} values of just the left and just the right cortex, Φ^{max}_{left} and Φ^{max}_{right}, by themselves. But from the intrinsic point of view, per the exclusion postulate, only the maximum of integrated information over this substrate exists for itself, that is, is a Whole.

To analyze what happens to the mind of a split-brain patient viewed through the lens of IIT, consider a futuristic version in which the blunt and irreversible action of the neurosurgeon's scalpel are replaced by a *subtle knife*, an advanced technology that allows the physician to delicately, progressively, and reversibly inactivate these thin axonal tendrils.

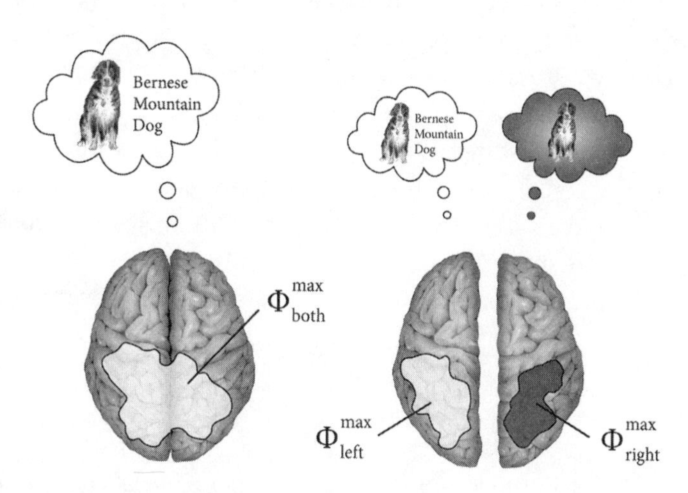

Figure 10.2

Splitting a brain into two minds: A single mind with its physical substrate, its Whole, spreads across both cortical hemispheres linked by the massive corpus callosum (left) in all of us. Its surgical disconnection creates two distinct minds with two Wholes (right), one that can speak and a second one that is linguistically incompetent. Neither mind has any direct acquaintance of the other, believing itself to be the sole occupant of the skull.

As callosal axons are turned off, one by one, the interhemispheric communication bandwidth decreases. Assuming the patient is conscious during the operation, his feeling will not be dramatically affected at first; that is, he may not notice any change in his experience of the world and of himself. Concurrently, Φ^{max}_{both} will change only little.

There will come a moment when turning off a single additional axon with the subtle knife causes Φ^{max}_{both} to dip below the bigger of Φ^{max} values for the independent hemispheres, say, below Φ^{max}_{left}. Given the reduced interhemispheric spike traffic relative to the intrahemispheric connections, the single Whole that extends across both hemispheres abruptly split into two Wholes, one in the left and one in the right hemisphere (fig. 10.2).

The single mind vanishes and is replaced by two minds, one housed in the left, with Φ^{max}_{left}, and the other in the right hemisphere, with Φ^{max}_{right}. One mind, supported by a Whole in the left, accesses Broca's area in the left inferior frontal gyrus, and can name what it sees (the dog breed in figure 10.2). It is oblivious to the presence of the other mind, supported by a Whole in the right hemisphere. From an intrinsic perspective, the other mind may as well be on the dark side of the moon.[5]

By closely observing the action of its body, the left mind could infer that "there's someone in my head but it's not me," as Pink Floyd laments in "Brain Damage." Conflict can arise between what the left mind wants and what the left side of the body, controlled by the right mind, does. Patients unbutton their shirt with one hand while undoing this action with their other hand or complain that their hand is controlled by an outside presence, doing things on its own. These actions are initiated by a conscious mind unable to speak "its mind." Such hemispheric rivalry eventually ceases, with the left hemisphere establishing dominance over the entire body.[6]

Brain-Bridging and the Über-Mind

Now consider the inverse—rather than splitting one brain let's fuse two brains into one. Imagine a futuristic neurotechnology, *brain-bridging*, that safely reads and writes billions of individual neurons with millisecond precision. Brain-bridging senses the spiking activity of neurons in the cortex of one person and synaptically links this to neurons in the corresponding area in the cortex of another person and vice versa, acting as an artificial corpus callosum.

This situation may occur naturally when twins are born with conjoint skulls (craniopagus). In one such case, illustrated in a striking YouTube video, two girls happily run around, giggling and playing together, just like regular kids do, except that their skulls are attached to one another.[7] They are truly inseparable. It is possible that each twin may have access, at least some of the time, to what the other perceives, while still retaining her own distinct mind and personality.

Let us brain-bridge your and my cortex, starting with a few wires linking our visual cortices. When I look at the world, I see what I usually see but now with a ghostly image of what you see superimposed, as if I'm wearing augmented reality goggles. How vividly and what aspects of what you see I experience depends on the details of the cross-brain wiring. But as long as the integrated information of your brain, Φ^{max}_{you}, and of my brain, Φ^{max}_{me}, exceed Φ^{max}_{both} of our interconnected nervous systems, we will retain our distinct minds. You will still be you, and I will be me, a consequence of IIT's exclusion postulate. The Φ calculus considers every possible partition to evaluate how irreducible the circuit is, and as long as the interbrain pathways are dwarfed by the massive, existing intrabrain pathways, nothing too dramatic changes.

As the bandwidth of the brain-bridge further cranks up—as more and more neurons are interconnected (probably on the order of tens of millions)—there will come a point when Φ^{max}_{both} exceeds either Φ^{max}_{you} or Φ^{max}_{me}, even if only by a little. At that moment, your conscious experience of the world vanishes, as does mine. From your and my intrinsic perspective, we cease to exist. But our death coincides with the birth of a new amalgamated or blended mind. It has a Whole extending across two brains and four cortical hemispheres (fig. 10.3). It sees the world through four eyes, hears through four ears, speaks with two mouths, controls four arms and legs, and shares the personal memories of two lives.

For illustrative purposes I assume in figure 10.3 that you are a native French speaker. When we are separate, we each see the dog and think of its breed in our mother tongue. The blended mind will have fluent access to both.

At times, I yearn to fully fuse with my wife's mind, to experience what she experiences. Sexual union only dimly and transiently achieves this desire—though our bodies are entangled, our two minds retain their separate identities. Mind blending via brain-bridging would allow transcendence, a complete union, an orgasmic dissolution of our distinct identities and the birth of a new soul.

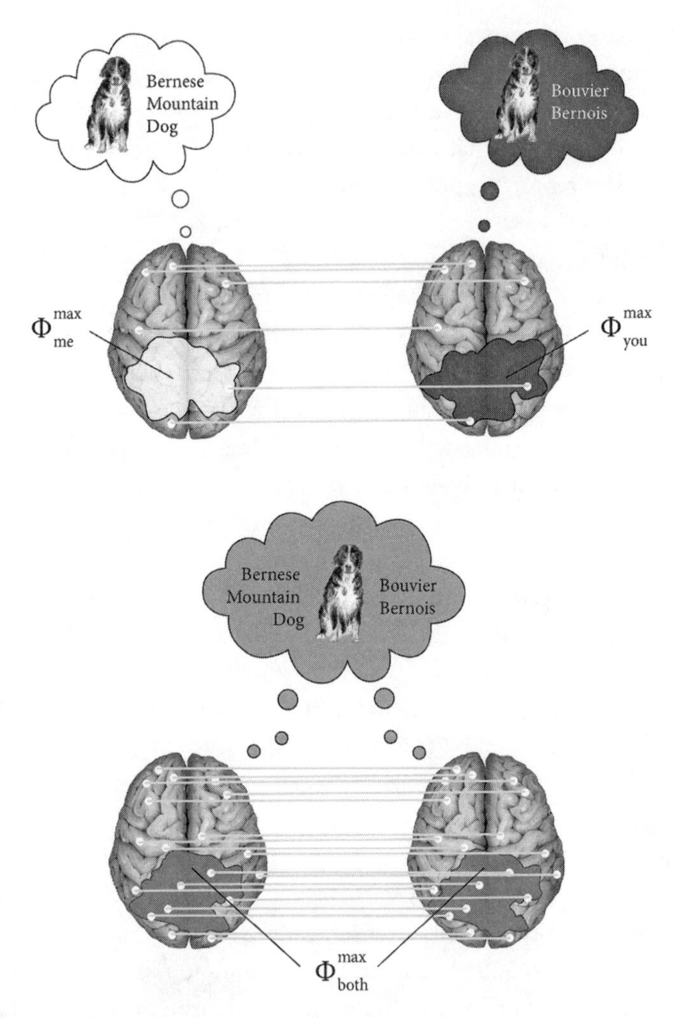

Figure 10.3

Melding two brains into one mind: Two brains are connected via brain-bridging, a yet-to-be-invented technology that would permit neurons in two brains to directly and reciprocally influence each other. In the top diagram, the effective connectivity of this artificial corpus callosum is assumed to be low, such that the integrated information within each brain, Φ^{max}_{me} and Φ^{max}_{you}, is bigger than the integrated information across both. While each mind has access to some information from the other brain, they retain their distinct identities (including in this example, thinking in English and in French). This changes radically when the number of neurons in the two brains that are interconnected exceeds some threshold (bottom). Abruptly, a single consciousness comes into being, with a single Whole extending across two brains.

Richard Wagner's opera, *Tristan und Isolde*, prefigures my thought experiment by more than a century. To the soundscape of the most rapturous music ever composed, Tristan yearns to be Isolde, exclaiming "Tristan you, I Isolde, no longer Tristan," to which Isolde echoes "You Isolde, Tristan I, no longer Isolde," transitioning to an ecstatic duet "Un-named, free from parting, new perception, new enkindling; ever endless self-knowing; warmly glowing heart, love's utmost joy!" If we are to believe Wagner, who himself wrote the libretto, mind-blending is bliss without end.

In practice, an abrupt comingling of two minds that developed autonomously over decades might lead to massive conflicts and pathologies, could end catastrophically, and would surely lead to a new specialty of blended-mind psychiatry.

Mind blending is fully reversible—as the bandwidth of the brain-bridge is dialed down, the blended mind disappears when the integrated information of the connected brains dips below the integrated information of one of them. You and I find ourselves back in our accustomed individual minds. Conceptually, cutting the link between our two brains is similar to split-brain surgery.

I see no reason for bridging to be limited to two brains. With sufficient advanced technology, it should be possible to link three, four, or hundreds of brains, limited only by the laws of physics. As each brain merges with the Whole, it adds its unique capabilities, intelligence, memories, and skills to the growing über-mind.

I foresee cults and religious movements springing up around über-mind bridging in a quest to abandon individuality in service of a greater Whole. Will this group mind acquire new mental powers beyond those of individuals, echoes of the Overmind in Arthur C. Clarke's *Childhood's End*? Is this the supreme destiny of humankind, a singular consciousness encapsulating the mental essence of our species?

A more terrifying depiction of such a group mind is the Borg, the fictional alien species from the *Star Trek* universe. It relentlessly absorbs sentient creatures into its hive mind ("All resistance is futile"). In the process, they give up their identity to the collective.

Given our current primitive ability to read, let alone write (stimulate), large numbers of individual nerve cells, human brain-bridging lies in the hazy future, but doing this in mice is more feasible.[8]

Neuronal Hegemony and Multiple Minds

Let me discuss another prediction of IIT. Per the exclusion postulate, of all the various circuits that include the substrate, only the one with the most integrated information exists for itself, is a Whole. As long as it controls, directly or indirectly, the speech area, it can speak its mind. This is the experiencing "I."

But why should there be only a single Whole across the 16 billion cortical neurons and their auxiliaries in the thalamus, basal ganglia, claustrum, amygdala, and other subcortical regions? As long as the substrates of these local maxima of Φ do not overlap, they can coexist independently and simultaneously within a single brain, each with its own definite borders.

A priori, there is no formal, mathematical reason for there to be a only single Whole that extends its hegemony across the entire cortical realm. In principle, there could be many nonoverlapping Wholes, like nations sharing a continent, each with its own experience.

History teaches that empires fall apart if countries or tribes at their distant periphery are only weakly controlled by the center. And so it may be with cortical networks. That is, without a strong central control there could always be a multitude of Wholes in a single brain. Is there a *hegemony* principle at work in the brain that counteracts these centrifugal forces? Could a single large Whole be overwhelmingly more likely than several smaller ones in networks with certain structural features (clustering coefficients, degree, path length)? Alternatively, specialized networks, such as the claustrum I briefly mentioned in chapter 6, may be responsible for coordinating the responses of large coalitions of cortical neurons.[9]

IIT envisions the possibility that the brain may condense into a large Whole and one or more smaller ones.[10] For instance, in a normal brain there might be one dominant Whole in the left hemisphere that peacefully coexists, under certain conditions, with a separate Whole within the right hemisphere. Their footprint may be dynamic, shifting, expanding and contracting, depending on the excitatory and inhibitory interactions among the neurons at different time scales. Each one may be specialized for a particular perceptual, motor, and cognitive task. Each one with its own experience, even though only the one in control of Broca's area can talk. The existence of two Wholes could explain many otherwise puzzling phenomena.

Consider a mild form of detachment from reality known as *mind wandering*. It happens whenever your attention drifts from the task at hand.[11] It is more common than you think—daydreaming while doing the laundry or preparing dinner, or listening to a podcast while driving. Part of your brain is processing the visual scene and adjusting the steering wheel and gas pedal appropriately, while your experiential self is far away, following the storyline.

The conventional interpretation is that driving is such a routine aspect of life that your brain has wired up an unconscious, zombie circuit. However, an alternate and more unorthodox explanation is that the sensorimotor and cognitive activities are each supported by their own Whole. Each has its own stream of consciousness, with the crucial difference being that only the Whole in charge of Broca's area can speak about its experiences. The other is mute and may not even be able to lay down a memory trace that could be accessed later on. As soon as something occurs that draws your attention, say the red break lights of the truck ahead abruptly turning on, the smaller Whole merges with the bigger one, and your brain quickly returns to a single mind. It is no easy matter to look for a smaller Whole in the brain, but some have devised ingenious ways to search for its hidden signatures.[12]

Another case of multiple Wholes might occur during dissociations known as *conversion disorders* (including what used to be called *hysteria*). The individual acts blind, deaf, or paralyzed, with all the relevant behaviors and manifestations, yet without any organic substrate. No stroke, injury, or other cause explains their symptoms that are clearly distressing to the patient. Their nervous system looks to be working just fine. More extreme forms are the loss of self and memory (psychogenic *fugue*) and the fragmentation of self into multiple separate streams of consciousness with separate perceptual abilities, memories, and habits (*dissociative identity disorder*). Historically, these cases have been analyzed on the psychoanalyst's couch or in the psychiatric ward. But the wide variety of dissociative symptoms might be more profitably explained by a network analysis based on functional and dysfunctional connectivity.[13]

Brains whose architecture differs widely from mammalian brains, such as the nervous systems of insects or cephalopods that are dominated not by a single, large cephalic circuit but by many ganglia distributed throughout their bodies, may operate continuously in such a multiple mind mode.[14]

Pure Experience and the Silent Cortex

Every one of our waking moments is filled with something—we travel into the future, recall our past, or fantasize about sex. Our mind is never still but constantly flutters hither and yon. Many of us dread being left alone with our mind and immediately whip out our smartphones to prevent such disquieting moments.[15]

It seems that any experience has to be about something. Does it even make conceptual sense to be conscious without being conscious of something? Can there be an experience not involving seeing, hearing, remembering, feeling, thinking, wanting, or dreading? What would it feel like? Would such a putative state of pure consciousness have any phenomenological attribute that distinguished it from deep sleep or death?

The quieting of the mind, its stilling until it is suspended in a divine nothingness, is a quest common to many religious and meditative traditions throughout the ages. Indeed, the core of so-called *mystical experiences* is the perfectly tranquil mind, cleansed of any attribute—*pure experience*. Within Christianity, Meister Eckhart, Angelus Silesius, a priest, physician, poet, and contemporary of Descartes, and the anonymous author of the fourteenth-century *The Cloud of Unknowing* all refer to such moments. Hinduism and Buddhism have a variety of related ideas, including "pure presence" or "naked awareness," and developed meditation techniques to attain and maintain such states. The eighth-century Tibetan Buddhist master, Padmasambhava, also known as Guru Rinpoche, writes:

> And when you look into yourself in this way, nakedly (without any discursive thoughts) since there is only this pure observing, there will be found a lucid clarity without anyone being there who is the observer; only a naked manifest awareness is present.[16]

These different traditions emphasize a state of void, the complete cessation of all mental content.[17] Awareness is vividly present yet without any perceptual form, without thought. The mind as an empty mirror, beyond the ever-changing percepts of life, beyond ego, beyond hope, and beyond fear.[18]

Mystical experiences often mark a turning point in the life of the experiencer. Recognizing this constant aspect of mind, divorced from perceptual and cognitive processes, leads to lasting emotional equanimity and increases well-being.[19] Subjects speak of having glimpsed a lucid, stable quality of mind that is indescribable.

Inhaling the rapid, short-acting, and powerful psychedelic N,N-Dimethyltryptamine (DMT), also known as the spirit molecule, can lead to similar mystic states, as does awakening from a near-death experience on the operating table or at the scene of a traffic accident. A safer alternative is a sensory deprivation tank.

I recently took the plunge and visited an *isolation* or *flotation tank* in Singapore, where my daughter lives. We each had our own pod, filled with water heated to body temperature. I stripped naked and entered the water, floating as if in the Dead Sea, as 600 pounds of Epson salt were dissolved in the bathtub. Once I closed the cover of this pod, housed in its own room, it became its own space capsule, utterly dark and quiet, except (after a while) for my beating heart.

It took me some time to become accustomed to and comfortable with this situation, time during which I lost any sense of where my body was situated in space. Once I emptied my mind of the daily residue of images and silent speech, I sank ever deeper into a bottomless, dark pool, suspended in a sightless, soundless, odorless, bodiless, timeless, egoless, and mindless space.

After several hours, my daughter became concerned enough at my silence to call out. This brought me back to everyday reality with all its clutter and unbidden images and voices, aches and desires, concerns and plans.[20] However, for a timeless moment, I apprehended something extraordinary, immensely valuable—a coming home to a state of pure being.[21]

It is difficult to say anything more about this ineffable state. In these troubled times, I frequently take myself back in my mind to this dark pool, to try to recapture my sense of equanimity.

Mystical experience are not paranormal (or parapsychological) events, even though they are sometimes classified as such. They are veridical experiences that arise in a natural way from the brain.

The existence of pure experience is anathema to contemporary cognitive psychology that is functionally grounded.[22] On this approach, something as pronounced as an experience must have one or more functions that promote the survival of the species. Yet when experiencing a state of pure consciousness, what function is being performed? Sitting immobile in meditation or floating in the dark and quiet water, without any inner speech, memory stream, or mind wandering, no computation is going on in any conventional sense. No sensory input is analyzed, nothing in the environment changes, nothing needs to be predicted or updated.

IIT, though, is not about performing any function. It is not a theory of information processing. Indeed, I never mentioned the specific function carried out by the binary network in figure 8.2. I only considered its intrinsic causal powers.

Nowhere does the theory require that information be broadcast throughout the brain for consciousness to occur. The maximally irreducible cause-effect structure associated with a Whole depends not only on the connectivity among its neurons and their internal mapping but also on their current state, that is, whether they are active (spiking) or not. Importantly, elements that are turned off, that are inactive, can selectively constrain past and future states of the system just as well as active ones. That means that inactive elements, neurons that don't fire, can still contribute intrinsic causal power. A denial not issued, a deadline that passed without the threatened consequence, a critical letter not sent can be consequential events, with their own causal power.

Seemingly paradoxically, a system can therefore be (nearly) silent (with only background activity), with all units off, yet still have a maximally irreducible cause-effect structure with some integrated information. At least in principle, a quiet cortex generates a subjective state, just as an agitated cortex does.

This runs profoundly counter to the professional instinct of neuroscientists—we spend much of our time devising ever-more sophisticated ways to track neural activity with microelectrodes, microscopes, EEG electrodes, magnetic scanners, and so on. We measure its amplitude and statistical significance and relate it to perception and action. That the absence of activity (relative to some background level) might constitute an experience is difficult to swallow.

How would a nearly silent cortex feel? Without significant activity, the mind experiences no sounds, no sights, no memories. It has one distinction, constituted by nonfiring neurons. The brain in this state has intrinsic causal powers, unlike the brain during deep sleep and its attendant loss of consciousness.[23]

An Inactive versus an Inactivated Cortex

This brings me to another counterintuitive prediction. At the heart of IIT is its focus on intrinsic, irreducible cause-effect power: the difference that makes a difference to itself. The more irreducible the cause-effect power, the more the system exists. Thus, any reduction in a system's ability to be influenced by its past and to determine its future will decrease its causal powers.

Let us stay with the nearly silent posterior hot zone of a meditator experiencing no content, with the associated pyramidal neurons firing little. In a thought experiment, we perfuse the drug tetrodotoxin, a potent neurotoxin found in the pufferfish (fugu), into the posterior hot zone. This chemical compound prevents neurons from generating any action potentials, shutting down their activity.

From the point of view of spiking activity, the situation changes little from before. In neither case are posterior cortical neurons active. In the original situation, they could be active but aren't, while in the latter case, the neurons aren't active because they have been drugged.

On a conventional reading of neuroscience, the phenomenology of the meditator is the same in both situations, pure experience, as from the point of view of the targets of these neurons, there is no difference—no spikes are leaving that region of cortex. Yet according to integrated information theory, the reduction of causal power makes all the difference and the two situations differ profoundly: pure consciousness in one and unconsciousness in the other.

To an outsider observing these neurons, both situations are alike. In neither case are neurons active. How can these two situations be so different with respect to experience? Well, first, the physical state of cortex is not the same in the two cases, for in the latter, specific ionic channels are chemically blocked.[24] Second, in IIT, the information is not in the message that is broadcast by neurons, but in the form of the cause-effect structure specified by the Whole. Neurons that could fire but do not also contribute to determining the causes and effects of any one state. Recall the dog that did not bark in the Sherlock Holmes story I mentioned in chapter 2. If these neurons become causally impotent by the action of the neurotoxin, they cease to contribute to consciousness.

IIT offers a rich bounty of unexpected predictions, such as the über-mind, pure consciousness, and the profound difference between an inactive and an inactivated cortex. I look forward to the next decade of experimental efforts seeking to validate these predictions.

I have alluded several times to the function of consciousness. What is it? How can IIT account for why experiences evolved? Read on.

11 Does Consciousness Have a Function?

"Nothing in biology makes sense except in the light of evolution" is a famous saying by the geneticist Theodosius Dobzhansky. Any aspect of an organism, whether an anatomical feature or a cognitive capacity, must convey some selective advantage to the species, or must have done so in the past. Considered in this light, what is the adaptive advantage of subjective experience?

I will start out by reminding you about the remarkable range of things we can do without being conscious. This raises the question of whether experience has any adaptive function at all. I then discuss an *in silico* evolutionary game in which simple animats, over tens of thousands of generations of births and deaths, become adapted to their environment. As they evolve, their brains become more complex, with ever-increasing integrated information. I end the chapter by taking up, again, the important distinction between intelligence and consciousness, between being smart and being conscious.

Unconscious Behaviors Rule Much of Our Life

Over the years, scholars have proposed a diverse welter of functions for consciousness. These range from short-term memory, language, decision making, planning actions, setting long-term goals, error detection, self-monitoring, inferring the intentions of others, to humor.[1] None of these hypotheses has found acceptance.

Despite humanity's brightest minds obsessing about the function of the sensitive soul (in Aristotelian parlance), we still do not know what survival value is attached to experience. Why are we not zombies, doing everything we do but without any inner life? On the face of it, nothing in the laws of physics would be violated if we didn't see, hear, love, or hate but still acted as we did. But here we are, experiencing the pains and pleasures of life.

The mystery of the function of consciousness deepens with the realization that much of the ebb and flow of life occurs beyond the pale of consciousness. A prima facie example are the well-rehearsed sensory-cognitive-motor actions that make up the daily routine of life. Crick and I called these *zombie agents* (see chapter 2), Bicycling, driving, playing a violin, running on a rocky trail, rapidly typing on a smartphone, navigating a computer desktop, and so on—we perform all these actions with nary a thought. Indeed, the smooth execution of these tasks requires that we do not focus too much on any one component. Whereas consciousness is needed to acquire and reinforce these skills, the point of training is to free up the mind to focus on higher-level aspects—such as the content of what we want to text or the looming thunderclouds during a climb—and to trust to the wisdom of the mind–body and its unconscious controller.[2]

We become experts and develop intuitions for these skills, the uncanny ability to execute the right move or to know the right answer without knowing why. Most of us have only a weak grasp of the formal syntactical rules underlying language. Yet we have no trouble intuiting whether or not a particular sentence in our native language is correctly formulated, without being able to explain why. Professional chess or Go players take one look at a board position and instinctively know the lines of attack and defense, even if they can't always fully articulate their reasoning.[3]

Yet we only develop expertise for a narrow range of capabilities, depending on our idiosyncratic interests and needs. As a consequence, we frequently need to solve new problems on the fly, without having encountered them previously. This would certainly seem to demand consciousness. Yet even here a cottage industry of psychologists argues that complex mental tasks, including adding numbers, making a decision, understanding who does what to whom in a painting or photo, and so on can be performed subliminally, without any awareness.[4]

Some take these experiments to their limits and argue that consciousness has no causal role at all. They accept the reality of consciousness but argue that feelings have no function—they are froth on the ocean of behavior, without consequence to the world. The technical term is *epiphenomenal*. Think of the noise that the heart makes as it pumps blood: the cardiologist listens to these beats with a stethoscope to diagnose heart health, but the sounds themselves are irrelevant to the body.

I find this line of argument implausible. Just because consciousness isn't needed to accomplish a well-rehearsed and simplistic laboratory task in no way implies that consciousness has no function in real life. One similar grounds, it could be asserted that legs and eyes have no function because somebody with his legs tied can still hop around and a blindfolded person can still orient herself in space. Consciousness is filled with highly structured percepts and memories of sometimes unbearable intensity. How could evolution have favored such a tight and consistent link between neural activity and consciousness if the feeling part of this partnership had no consequences for the survival of the organism? Brains are the product of a selection process that has operated over hundreds of millions of birth-and-death cycles. If experience had no function, it would not have survived this ruthless vetting process.

We also need to consider the likelihood that experience could be a by-product of the selection of other characteristics, such as highly flexible and adaptive behavior, rather than being directly selected for. In evolutionary language this is known as a *spandrel*. Popular examples of spandrels are humanity's wide-spread love of music or the ability to engage in higher mathematics. It is likely that neither music appreciation nor math skills were directly selected for in hominid evolution, but that they emerged when big brains made these activities possible. And so it could be with consciousness.

Integrated Information Is Adaptive

Integrated information theory takes no position on the function of experience as such. Any Whole feels like something; it doesn't even have to do anything useful for it to have an experience. Indeed, you may have noticed that I never mentioned the input–output function of the simple circuit in figure 8.2. Furthermore, as I discussed in the previous chapter, a nearly silent cortex may be the substrate of a pure experience, without any ongoing information processing.

In this strict sense, experience has no function. This situation is analogous to that of physics having nothing to say about the utility of mass or electrical charge. Physicists don't worry about the "function" of either. Rather, in this universe, mass describes how spacetime is curved and how charged particles are repelled and attracted to each other. Agglomerations of particles, such as proteins, have a net mass and charge that influences their dynamics and propensity to interact with each other. This determines their behavior, upon which evolution acts. That is, while mass and charge have no function in a strict sense, they help to shape function in a broader sense. And so it is with intrinsic cause-effect power.

IIT provides an elegant account of why conscious brains evolved. The world is immensely complex, across many spatial and temporal scales. There is the physical environment with its caves and burrows, forests, deserts, daily weather and annual seasons, complemented by the social environment of prey and predator, potential mates and allies, each with their own motivation that organisms need to infer and keep track of.

Brains that incorporate the associated statistical regularities (e.g., antelopes usually arrive at the drinking hole just after sunset) into their own causal structure have an edge over brains that don't. The more we know about the world, the more we are likely to survive.

A number of colleagues and I set out to demonstrate this by simulating the evolution of digital organisms over eons of time and tracking how the integrated information in their brains changes as they adapt to their surroundings.[5] This is known as *in silico* evolution, as in the video games SimLife or Spore.

The simulated creatures, animats, are endowed with a primitive eye, a proximity sensor, and two wheels. A neural network, whose connectivity is genetically specified, links the sensors to the motors. Animats survive by navigating as quickly as possible through a two-dimensional maze. At the start of an evolutionary run, their connectome, or map of neural connections, begins as a blank slate. Three hundred of these animats, each one slightly different from the others, are placed into a labyrinth to see who can navigate the farthest. Initially, most just stumble around, turn in circles or remain totally immobile. A few might move in the right direction, even if by only one or a few steps.

Animats have a fixed lifespan, at the end of which the thirty best-performing ones are selected to seed the next generation of three hundred animats. Each newborn generation comes with some slight, random variation in its genetic code specifying the layout of its brain, as this is the raw material on which natural selection operates on. Hopefully, a few of them will move a bit farther along than their parents did. After 60,000 generations of life and death, the distant descendants of the animats that blindly stumbled about at the creation of their world have become highly adept, quickly traversing each maze they encounter.[6] This game is replayed again and again—simulating different evolutionary trajectories that never repeat themselves. Each evolutionary game gives rise to endless forms of animats, each one with its own distinct nervous system, echoing Darwin's famous closing sentence in his *On the Origin of Species*:

> There is grandeur in this view of life, with its several powers, having been originally breathed into a few forms or into one; and that, whilst this planet has gone cycling on according to the fixed law of gravity, from so simple a beginning endless forms most beautiful and most wonderful have been, and are being, evolved.

When the integrated information of the brains of animats from different points along their evolutionary lineage is plotted against how well and fast they traverses mazes the result is clear and compelling (fig. 11.1): a positive relationship between how well an organism is adapted and its Φ^{max} emerges. The more integrated its brain, the more irreducible its neural network linking its input to its output, the better it does.

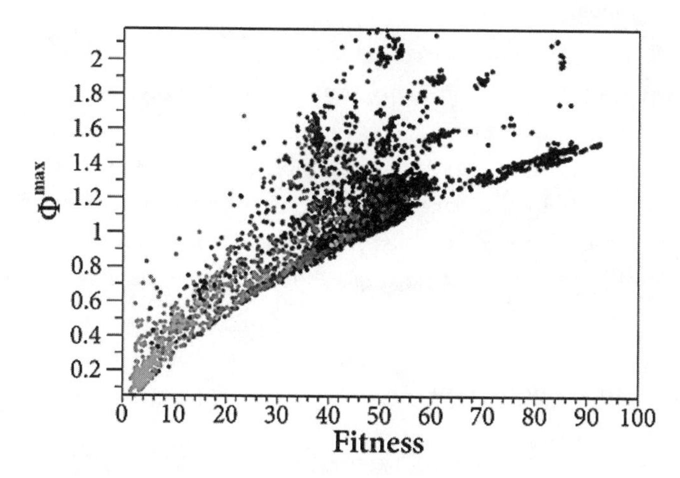

Figure 11.1

Evolving integrated brains: As digital organisms evolve to run mazes more efficiently, the integrated information of their brains increases. That is, increasing fitness is associated with higher levels of consciousness. (Modified from Joshi, Tononi, & Koch, 2013. This study used an older version of IIT, in which integrated information is computed somewhat differently from Φ^{max} in the current formulation.)

Particularly striking about the figure is the existence of a minimum of Φ^{max} for any one fitness level. Once this minimal integration has been achieved, organisms can acquire additional complexity without altering their fitness. So in this broader sense, experience is adaptive: it has survival value.

A different genus of animats, evolved to catch falling blocks as in the game of Tetris, shows similar trends. As adaptation increases, so do the animats integrated information and the number of distinctions the system is capable of supporting.[7] Thus, evolution selects organisms with high Φ^{max} because, given constraints on the number of elements and connections, they pack more functions per element than their less integrated competitors; they are more adept at exploiting regularities in a rich environment.

Extrapolated from the action of tiny animats to humans, this view is broadly compatible with the executive summary hypothesis formulated by Francis Crick and myself:

> Our… assumption is based on the broad idea of the biological usefulness of visual awareness (or, strictly, of its neural correlate). This is to produce the best current interpretation of the visual scene, in the light of past experience either of ourselves or of our ancestors (embodied in our genes), and to make it available, for a sufficient time, to the parts of the brain that contemplate, plan and execute voluntary motor outputs.[8]

Any one conscious experience contains a compact summary of what is most important to the situation at hand, similar to what a president, general, or CEO receives during a briefing. This precis enables the mind to call up relevant memories, consider multiple scenarios, and ultimately execute one of them. The underlying planning occurs largely nonconsciously, as that's the responsibility of the unconscious homunculus (or, in this metaphor, the responsibilities of the staff reporting to the executive), largely confined to prefrontal cortex (see chapter 6).

The Intelligence–Consciousness Plane

This brings me to a general observation about the relation between intelligence and consciousness.

IIT is not concerned with cognitive processing as such. It is not about attentional selection, object recognition, face identification, the generation and analysis of linguistic utterances, or information processing. IIT is not a theory of intelligent behavior in the same sense that the theory of electromagnetism is not a theory of electric machines but of electromagnetic fields. The associated Maxwell equations are, of course, pregnant with consequences for prospective motors, turbines, relays, and transformers. And so it is with IIT and intelligence.

What these *in silico* evolution experiments demonstrate is that adapted organisms possess a degree of integrated information reflecting the complexity of the habitat they've adapted to. As the diversity and richness of these niches grow, so do the nervous systems exploiting the attendant resources as well as their intrinsic causal powers—from a few hundred neurons in tiny worms, to a hundred thousand in flies, to a hundred million in rodents, to a hundred billion in humans.

Commensurate with this increase in brain size is the growing ability of these species to learn to deal with novel situations. Not by instinct, which is another word for innate, hardwired behaviors—freshly hatched sea turtles seeking the protection of the sea, or bees instinctively knowing how to dance to convey the location and quality of a food source—but by dint of learning from previous experiences, as when a dog learns that the kibble is in this cupboard and that the garden is accessed through that door. We call this ability intelligence. By this measure, bees may have less intelligence than mice whose cortex conveys flexibility to readily learn certain contingencies, dogs are smarter than mice, and we are smarter than our canine companions.

People differ in their ability to understand new ideas, to adapt to new environments, to learn from experience, to think abstractly, to plan, and to reason. Psychologists capture these differences in mental capabilities via a number of closely related psychometric concepts and measures such as general intelligence (g, or general cognitive ability), and fluid and crystalline intelligence. Differences in people's ability to figure things out quickly and to apply insights that they learned in the past to current circumstances are assessed by psychometric intelligence tests. These are reliable, in that different tests strongly correlate with one another. They are also stable across decades. That is, measures such as the intelligence quotient (IQ) can be repeatedly and reliably obtained from the same subjects nearly seventy years later. Animal behaviorists have defined the mouse equivalent of this human g-factor.[9] So, for the sake of the following argument I assume the existence of a generalized, cross-species, single intelligence factor, big G.

Intelligence ultimately is about learned, flexible actions. All else being equal, for instance, in the degree of sophistication of individual nerve cells, nervous systems with more neurons should have more elaborate and flexible behaviors than brains with fewer cells and, therefore, a higher G factor. However, given our very limited understanding of the neural roots of intelligence, the relationship is probably considerably more intricate.[10]

How does brain size affect consciousness? Bigger networks have combinatorially more potential states than smaller ones. Of course, that is not a guarantee that integrated information and causal power likewise increases with network size (think of the cautionary tale of the cerebellum), for that requires a balancing act between the opposing trends of differentiation and integration. However, it is fair to say that integrated information in nervous systems shaped by the fierce forces of natural selection over eons will increase with brain size. As a consequence, the ability of the current state of such a network to constrain trillions of its own past and future states becomes more refined as network size grows. That is, the bigger the brain, the more complex its maximally irreducible cause-effect structure may be, the bigger its Φ^{max} and the more conscious it becomes.[11]

What this means is that a large-brain species is not only capable of more phenomenal distinctions than a smaller-brained one (say, it could experience the hues of a billion distinct colors versus seeing the world only with a palette of thousands of colors, or it could sense the magnetic field or infrared radiation) but can also access more higher-order distinctions and relations (for instance, relating to insight and self-consciousness or to feelings of symmetry, beauty, numbers, justice, and other abstract notions).

Let me combine the different strands of this argument into a speculative graphical cartoon, the I-C plane (for intelligence-consciousness), of how smart species are versus how conscious they are. Figure 11.2 ranks five species, Medusa jellyfish with a loosely organized neural net, bees, mice, dogs, and people, according to my hypothetical G measure of intelligence and integrated information. Note that neither has a natural upper limit.

The plot highlights a monotonic relationship between intelligence and consciousness across species. Creatures with bigger brains are smarter as well as more conscious than species with smaller brains. More conscious within the context of IIT means more intrinsic, irreducible cause-effect power, more distinctions and relations. That same trend might also hold when comparing individuals within any one species.[12]

Exceptions to this relationship linking intelligence and consciousness are cerebral organoids. These are three-dimensional cellular ensembles derived

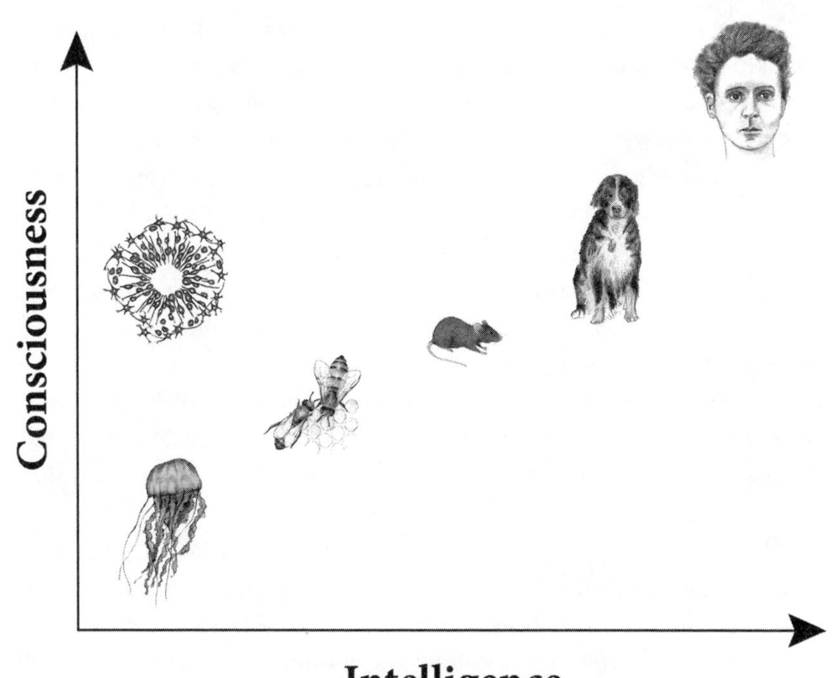

Figure 11.2

Intelligence and consciousness and their co-variation with brain size: Five species with nervous systems spanning eight orders of magnitude in number of neurons, from jellyfish to Marie Curie, standing in for humanity, are arrayed onto the intelligence-consciousness plane. Intelligence is operationalized as the ability to learn to respond flexibly to a constantly changing environment, and consciousness is measured as integrated information. As brain size increases, so do both intelligence and Φ^{max}. Such a diagonal trend is a hallmark of evolution by natural selection. This relationship can break down in engineered systems, such as cortical organoids (upper left) that have negligible intelligence but may have high Φ^{max}.

from human-inducible pluripotent stem cells that allow neuroscientists, clinicians, and engineers to grow tissue using a handful of mature starter cells from an individual child or adult. Reprogrammed by a quartet of "magic" transcription factors, they subsequently differentiate and self-organize in incubators.[13] It takes many months for these cells to mature into electrically active cortical neurons and their supporting glial cells, about the same as it takes to grow a human fetus from a fertilized egg. Organoids carry immense therapeutic promise for helping to understand neurological and psychiatric diseases.[14]

During infancy, the structured sensory input streaming from eyes, ears, and skin, in combination with appropriate wiggling of eyes, head, fingers, and toes that provides feedback signals, imposes order onto the immature nervous system of the baby with the help of synaptic learning rules. That's how all of us learned the causal structures of the particular environment into which we were born. Organoids lack any of this. Once this hurdle is overcome and these organoids express sophisticated synaptic learning rules, artificial patterned external stimuli could be imposed onto the organoid culture with dense arrays of computer-controlled electrodes, mimicking a crude and primitive developmental process.

Given the astonishing pace of progress in stem cell biology and tissue engineering, bioengineers will soon be able to grow sheets of cortical-like tissue on an industrial scale in vats, properly vascularized and supplied with oxygen and metabolites to maintain healthy levels of neuronal activity. These cortical carpets, woven far finer than any Persian rug, will have some integrated information with an accompanying irreducible Whole. The chance that it experiences anything like what a person feels—distress, boredom, or a cacophony of sensory impressions—is remote. But it will feel something. Therefore, to avoid the attendant ethical dilemma posed, it would be best if this tissue were anesthetized.[15]

But one thing is certain. Given that these organoids have no traditional sensory input or motor output, they won't be able to act in the world; they will not be intelligent. Their situation is akin to a brain, dreaming of a vast empty space, inside a paralyzed and sleeping body. Consciousness without intelligence: this places organoids in the upper left-hand corner of the I-C plane.

What about other engineered systems, in particular programmable digital computers? Can they have experiences? Where do they sit in the I-C plane? Before I come to this, let me first explain the strengths and the weaknesses of the computational theory of the mind. It is based on the conjecture that consciousness is computable.

12 Consciousness and Computationalism

What do Rachael in the cult science-fiction movie *Blade Runner*, Samantha in the Hollywood comedy *Her*, Ava in the dark psychodrama *Ex Machina*, and Dolores in the TV series *Westworld* have in common? None was born to a woman, all have attractive female attributes, and all are the object of desire of their male protagonists, demonstrating that lust and love extend to the engineered.

A future where the boundary between carbon-evolved and silicon-built life becomes porous is approaching us at warp speed. With the advent of deep machine learning, voice technology has reached near-human capability, creating the helpful spirits of Apple's Siri, Microsoft's Cortana, Amazon's Alexa, and Google's Assistant. Their linguistic skills and social graces are improving so relentlessly that soon they will become indistinguishable from a real assistant—except that they will be endowed with perfect recall, poise, and patience, unlike any flesh-and-blood being. How much longer before somebody falls in love with the disembodied digital voice of their personal digital assistant?

Their siren voices are living proof of the narrative of our times—that our mind is software, running on the computer that is our brain. Consciousness is just a couple of clever hacks away. We are only meat machines, no better, and, increasingly, worse, than computers. According to the more triumphalist voices in the tech industry, we should revel in our soon-to-come obsolescence; we should be grateful that *Homo sapiens* will have served as a bridge between biology and the inevitable next step in evolution, superintelligence. Smart money in Silicon Valley thinks so, op-ed pieces proclaim it to be so, and sleek sci-fi flicks reinforce this poor man's Nietzschean ideology.

Mind-as-software is the dominant mythos of liquid modernity, of our hyper-individualized, globe-trotting, technology-worshipping culture. It is the one remaining mythos of an age that believes itself to be immune to mythology. An age whose elite is witnessing with incomprehension and indifference the dying struggle of the once all-powerful mythos that sustained the West for two millennia—Christianity.

I here use mythos (or myth) in the sense of the French anthropologist Claude Lévi-Strauss, as the collection of explicit and implicit, spoken and unspoken guiding beliefs, stories, rhetoric, and practices that provide meaning to any one culture.[1] Mind-as-software is an unspoken background assumption that needs no justification. It is as obvious as the existence of the devil used to be. For what is the alternative to mind-as-software? A soul? Come on!

In reality, though, mind-as-software and its twin, brain-as-computer, are convenient but poor tropes when it comes to subjective experience, an expression of functionalist ideology run amok. They are more rhetoric than science. Once we understand the mythos for what it is, we wake as from a dream and wonder how we ever came to believe in it. The mythos that life is nothing but an algorithm limits our spiritual horizon and devalues our perspective on life, experience, and the place of sentience in time's wide circuit.

Let us look under the hood of the mythos of computationalism to understand what it is and where it comes from.

Computationalism: The Dominant Faith of the Information Age

The zeitgeist of our age is that digital computers can ultimately replicate anything that humans can do. Therefore, they can also be anything that humans can be, including conscious. Note the subtle but critical shift from *doing* to *being.*

Computationalism or the computational theory of mind is the reigning doctrine of mind in Anglo-Saxon philosophy and computer science departments and in the tech industry. Its seeds were laid down more than three centuries ago by Gottfried Wilhelm Leibniz, whom you met in chapter 7. Leibniz was on a lifelong quest to develop a universal calculus, a *calculus ratiocinator.* He was looking for ways to cast any dispute into a rigorous mathematical form so that its truth could then be evaluated in an objective manner. As he wrote:

> The only way to rectify our reasonings is to make them as tangible as those of the Mathematicians, so that we can find our error at a glance, and when there are disputes among persons, we can simply say: Let us calculate, without further ado, to see who is right.[2]

Leibniz's dream of universal computation motivated logicians of the late nineteenth and early twentieth century, culminating in the 1930s with the work of Kurt Gödel, Alonzo Church, and Alan Turing. They constructed the foundation for the information age by two mathematical feats. First, their labors placed absolute and formal limits on what can be proven by mathematics, bringing its ancient, aspirational dream of formalizing truth, of constructing an *alethiometer*, a truth meter, to an end.[3] Second, they gave birth to the Turing machine, a dynamic model of how any computational procedure, no matter what, can be evaluated on an idealized machine.

The import of this intellectual feat is difficult to overemphasize. The Turing machine is a formal model of a computer, stripped down to its essentials. It requires four things: (1) an infinite tape for writing and storing symbols, such as 0 and 1, that serves as an input device and for storing intermediate results; (2) a scanning head that reads these symbols from the tape and can overwrite them; (3) a simple machine with a finite number of internal states; and (4) a set of instructions, a program really, that fully specifies what the machine does in each of these internal states—"If in state (100) and reading a 1 from the tape, switch into state (001) and move one square to the left," or "If in state (110) and reading a 0, remain in this state and write a 1." That's it. Anything that can be programmed on any digital computer, no matter whether it is a supercomputer or the latest smartphone, can, in principle, be computed by such a Turing machine (it may take a very long time, but that is a practical matter). Turing machines have attained such an iconic, foundational stature that the modern notion of what it means to compute is identified with "computable by such a machine" (the so-called Church–Turing thesis).

These abstract ideas about computability turned into room-filling electromechanical computing machinery, born in sin during World War II to improve artillery tables, design atomic weapons, and crack military ciphers. Carried by wave after wave of advances in solid state and optical physics, circuit miniaturization, mass production, and the forces of market capitalism (famously captured by Moore's law—the number of transistors in a dense integrated circuit doubles about every two years), digital computers radically upended society, the nature of work, and how we play. Less than a century later, the offspring of these gigantic machines with puny calculating abilities—the ENIAC, UNIVAC, Colossus, and so on of their day—pack powerful sensors and processor chips into a sleek, handheld glass and aluminum case. These are the intimate, personalized, and treasured artifacts we carry around everywhere and consult obsessively every few minutes. An amazing development that shows no sign of slowing down.

Artificial Intelligence and Functionalism

Modern AI is powered by two classes of machine-learning algorithms that originated in twentieth-century research on the neuroscience of vision and the psychology of learning.

The first is that of deep convolutional networks ("deep" refers to the large number of processing layers). These are trained by exposing them offline to vast databases of, say, labeled images of dog breeds or vacation pictures, financial loan applications, or French-English translated texts. Once taught in this manner, the software instantaneously distinguishes a Bernese mountain dog from a Saint Bernard, correctly labels vacation photos, identifies a fraudulent credit-card application, or translates Charles Baudelaire's "Là, tout n'est qu'ordre et beauté, Luxe, calme et volupté" into "There, everything is order and beauty, Luxury, calm and voluptuousness."

The mindless application of a simple learning rule turns these neural networks into sophisticated look-up tables with superhuman abilities.

The second class of algorithms uses reinforcement learning, dispensing altogether with human advice. This works best when there is a single goal that can be achieved by maximizing a numerical score such as in many board or video games. The software samples in a sophisticated manner the space of all possible moves in a simulated environment and chooses the action that maximizes the score. After playing four million games of Go against itself, DeepMind's AlphaGo Zero reached superhuman performance. It did so in hours, compared to the years of relentless training a gifted human needs to become a highly skilled Go master. Its descendant, such as AlphaZero, decisively ended the era of human dominance of classical board games. Algorithms now play Go, chess, checkers, many forms of poker, and video games such as Breakout or Space Invaders, better than any person. Because the software learns without the benefit of human intervention, many find this ominous and frightening.

Before any of these momentous developments took place, computers had already provided scholars with a powerful metaphor for how the brain operates—the computational or information-processing paradigm. In this narrative, the brain is a universal Turing machine—it transforms incoming sensory information to yield an internal representation of the external world. In conjunction with emotional and cognitive states and memory banks, the brain computes an appropriate response and triggers motor actions. We are Turing machines made flesh, robots unaware of our own programming.

Consider a common action: texting in response to something you just saw. Your retina acquires visual information at about one billion bits per second; this data stream is reduced a hundred-fold by the time the information leaves the eyeball. If you're nimble, you can type five characters per second, which, taking into account the entropy of English, is ten bits per second. Estimates for reading or speaking come up with approximately similar numbers. Somehow, the one trillion all-or-none spikes generated each second by your brain transform ten million bits of data streaming up the optic nerve into ten bits of motor information. And the same visuomotor system can be rapidly deployed to ride a bike, pick up seaweed with chopsticks, or compliment your friend on her new lipstick.[4]

Computationalism holds that your mind–brain does this as any Turing machine would—it performs a series of computations on the incoming data streams, extracts the symbolic information, accesses its memory bank, compiles everything into an answer, and generates the appropriate motor output.

On this view, the associated software, the mind, runs on a wet computer. Of course, the nervous system is not a conventional von Neumann computer—it operates in parallel, without a system-wide clock or bus, its elements switch at a glacial speed of milliseconds, memory and processing are not separate, and it uses mixed analog and digital signals—but it is a computer nonetheless. The details don't matter—so the argument goes; only the abstract operations implemented are relevant. And if the operations of this wet computer encased in bone are faithfully captured at the relevant level of representation by software executed on a silicon processor, then everything associated with these brain states, including subjective experience, will automatically follow. Nothing else is needed to explain consciousness.

Computationalism is a variant of functionalism, which holds that a mental state, such as a pleasurable experience, is independent of the internal constitution of the underlying physical mechanism. Any mental state depends only on the role it plays for the mechanism, including its relationship to the environment, sensory input, motor output, and other mental states. What matters, on this view, is the function of the mental state. The physics of the mechanism, the stuff out of which the system is made and how it is wired together, is irrelevant.

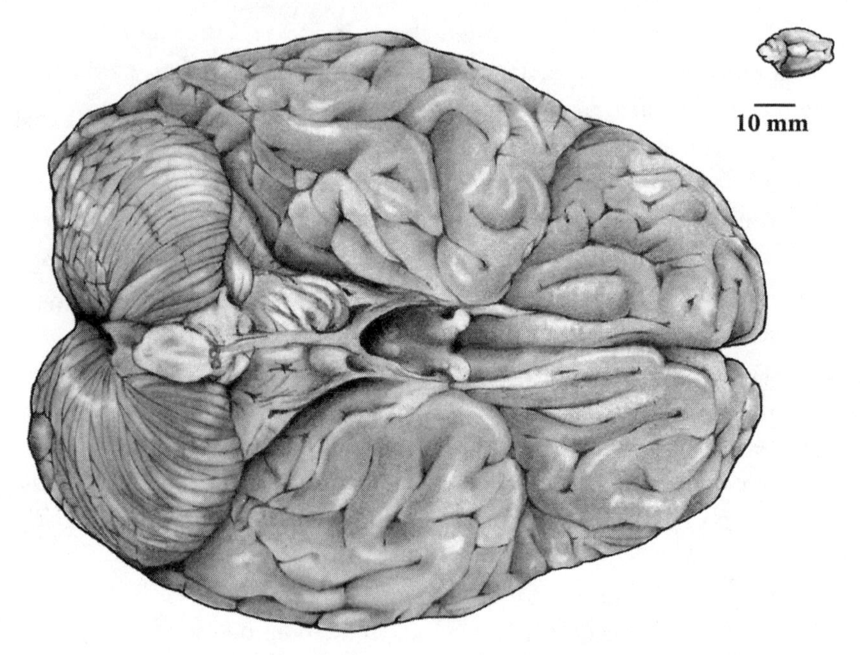

Figure 12.1
Computationalism: Today's dominant theory of mind argues that brains, here a human and a mouse brain drawn to scale, are nothing but wet approximations of Turing machines, with experience arising out of computation. This powerful mind-as-software metaphor has turned into an encompassing mythos for all life.

Some argue for a stricter criterion of functionalism. To have our experiences, computers should simulate not just our cognitive functions, but all the detailed causal interactions that occur in our brains, say at the level of individual neurons.[5]

On the Uses and Misuses of the Brain-as-Computer Metaphor

The poster child for the information-processing paradigm is the mammalian visual system. The stream of visual data ascends from the retina to a terminus in the first stage of cortical processing, primary visual cortex, at the back of the brain. From there onward, the data is distributed and analyzed in numerous cortical regions until it leads to perception and action.

Recording from the primary visual cortex of anesthetized cats, David Hubel and Torsten Wiesel at Harvard University described in the early 1960s a set of neurons they called "simple" cells.[6] Hubel and Wiesel would go on to win the Nobel Prize for their discoveries. These neurons responded to a tilted dark or light bar in a specific region of the visual field of the animal. Whereas simple cells are particular about where in visual space the oriented line is located, a second set of "complex" cells is less discerning about the exact location of that line. Hubel and Wiesel postulated a wiring scheme to explain their findings, consisting of multiple layers of cells—the first layer corresponds to the input cells that carry the visual information as captured by the eyes. These cells respond best to spots of light. They feed into a second layer of neurons, the simple cells, which talk in turn to a third layer of neurons, the complex cells.

Each cell is a processing element or unit that computes a weighted sum of its input and, if the sum is big enough, turns on the unit's output; otherwise, it remains off. The exact manner in which the units are wired up determines how cells in the input layer that respond to edges of any orientation are transformed into cells that care about a particular orientation at a particular location in the visual field. In a subsequent step, these cells provide input into units that discard some of that spatial information and signal an appropriately oriented line anywhere. Deep convolutional networks, the building blocks of the machine learning revolution, are direct descendants of these early cartoon models of the visual brain.

Subsequent discoveries of neurons in visual cortex that responded to faces reinforced these ideas—visual processing occurs within a hierarchy of processing stages in which the information flows upward, from units that care about primitive features such as brightness, orientation, and position to units that represent information in a more abstract manner, such as a generic woman's face or the specific face of Grandma or the actress Jennifer Aniston. This cascade of processing layers is called a *feedforward processing* (as in the cerebellar circuit). Each stage of processing influences only the next one down the line, not previous layers (which would be *feedback processing*).

Ironically, although machine learning networks are modeled on the brain, cortical networks are most certainly not feedforward circuits. Indeed, of all the synapses made among cortical neurons, only a minority, under one in ten, originate from connections from a previous stage of processing. The rest derive from nearby neurons or from cells in higher, more abstract processing stages feeding back onto earlier ones. Neural network theorists don't know how these massive feedback connections contribute to our ability to learn from a single example, something computers have trouble with.

Even though this textbook view of layered cortical processing, ascending, rung by rung, from primitive line-line features to more abstract ones, is being revised now that large-scale surveys of the visual responses of tens of thousands of cortical neurons are becoming available,[7] we can't help but interpret the way the brain works through the lens of this enormously successful feed-forward computing technology.

But many features of the nervous system defy a straightforward, brain-as-computer explanation.

Consider the retina—a finely woven piece of neural lace at the back of the eyes. About a quarter the size of a business card and not much thicker, it is structured like a Black Forest cake, with three layers of cell bodies separated by two layers of "filling," where all the synaptic and dendritic processing occurs. The incoming rain of photons is captured by a hundred million photoreceptors and turned into electrical signals that percolate through the various processing layers until it reaches about one million ganglion cells. Their output wires, the bundle of axons making up the optic nerve, convey spikes, the universal idiom of the nervous system, to their far-flung targets in the rest of the brain.

The computational job of the retina is straightforward—convert the light from a sun-drenched beach or a starlit nightscape into spikes. Yet for this seemingly simple task, biology needs about one hundred different types of neurons, each one with a unique morphology, a unique molecular signature, and a unique function. Why so many?[8] The image sensor in your smartphone does the same job with a handful of transistors underneath each pixel. What possible computational justification is there for employing this plethora of specialists?

The same abundance exists in cortex. Each cortical region has close to one hundred cell types. The inhibitory neurons are similar across the cortical sheet, while the excitatory neurons, in particular pyramidal neurons, differ between regions. This is likely because they send their information to different places and the zip code of these addresses is encoded in the genes of these neurons. Cell types have different cellular morphologies, neurotransmitters they are sensitive to, distinct electrical responses and so on. A brain is made up of upwards of a thousand or more types of cells.[9] Table 12.1 lists some of the other major architectural differences between evolved organisms and manufactured artifacts.

The computational metaphor is inadequate to explain this striking observation. Theory tells us that a combination of just two types of logic

Table 12.1
Differences between brains and computers

	Brains	Digital computers
Time	Asynchronous spike event	System-wide clock
Signals	Mixed analog–digital signals	Binary signals
Computations	Analog nonlinear summation followed by half-wave rectification and thresholding	Boolean operations
Memory	Tightly integrated with processors	Separation between memory and computation
Turing Universal	No	Yes
Types of computational nodes	Around 1,000	Handful
Speed of nodes	Millisecond (10^{-3} sec)	Nanosecond (10^{-9} sec)
Connectivity	1,000–50,000	<10
Robustness	Robust to component failure	Brittle

The radically different architectures of brains and digital computers makes all the difference when it comes to consciousness.

gates, expressing conjunction and negation (or variants thereof), is sufficient to instantiate any computation. Anything can be computed using enough AND and NOT gates. Digital computers get by with a handful of different types of transistors (including power transistors and specialized flip-flop circuits for solid-state memory).

Why this rococo exuberance of different brain cells? Does it have a computational function? My money is on betting that rather than subserving computational efficiency, cell types are the product of evolutionary, developmental, and metabolic constrains.[10]

Whole Brain Emulation

Even if we reject the computational view of the brain, there is no doubt that computers have the powerful ability to simulate the brain. Might this lead ultimately to a conscious mind?

Today, the principles underlying the operation of individual synapses, dendrites, axons, and neurons are reasonably well understood. The dynamics of these elements can be captured by nonlinear differential equations, variants of the famed Hodgkin–Huxley equations of action potential initiation and propagation.[11] Vast numbers of such equations, modified to account for synaptic interactions among neurons, have been run on supercomputers to simulate the spiking behavior of several hundreds of thousands neurons in a thin slice of rat cortex as part of Switzerland's Blue Brain Project. These simulate the dynamics of reverberant electrical activity in a brain slice.[12] Scaling such faithful models of networked neurons to encompass the entire brain of a mouse, with its 100 million cells, will become technologically possible in the next five years.

Yet such progress does not address the vastly more challenging problem of our inadequate knowledge of the prodigious complexity of the brain, from the molecular to the system level. Gazillions of parameters in these simulations need to be assigned specific values—channel densities, receptor binding concepts, coupling coefficients, concentrations, and so on. Without such detailed knowledge, neuroengineers can't breathe life into their simulacrum. Yes, they can get the software to do something that looks vaguely biological, but it'll be like a golem, stumbling about, trying to imitate a real brain. The dirty secret of computational neuroscience is that we still do not have a complete dynamic model of the nervous system of the worm *C. elegans*, though it only has 302 nerve cells and its wiring diagram, its connectome, is known. So here we are, trying to understand the human brain, when we do not yet understand the worm brain.

That is the deeper reason for why what AI enthusiasts refer to as *whole* (human) *brain* emulation lies decades in the future.[13] I state this with some conviction as I have dedicated much of my professional life to accurate simulations of neuronal circuits.[14] I will discuss in the next chapter whether such whole brain simulations would even be conscious.

Cultures view the mind–brain problem through the lens of whatever technology they are most familiar with. Plato and Aristotle imagined memory as writing on a wax tablet. Descartes conceived of animal spirits flowing through arteries, cerebral cavities, and nervous tubules in terms of the hydraulics that animated the moving statues of gods, satyrs, nymphs, and heroes of the fountains at the French court in Versailles. Subsequent metaphors likened the brain to a mechanical clockwork, a telephone switchboard, an electromechanical computer, the internet, and today, to deep convolutional networks or generative adversarial networks.

Curiously, computational metaphors are rarely applied to the liver or the heart. While scientists seek to build accurate computer models of these organs, they don't think of the metabolic processes in the liver or the pumping action of the heart in information-theoretical terms.

The dangers with metaphors is that we forget that they capture a limited aspect of reality. "All the world's a stage" is a fine, poetic figure of speech referring to some aspects of existence, but you and I are not hired actors, there is no audience, and no playwright gave us our lines.

The Global Neuronal Workspace Theory of Consciousness

Let me end this chapter by describing the computational view of consciousness, the central tenet or conjecture held by the digerati in academe and media. It is best encapsulated by the *global neuronal workspace* model of consciousness.[15]

Its lineage can be traced to the *blackboard architecture* of the early days of artificial intelligence, in which specialized programs access a shared repository of information, the blackboard or central workspace. The cognitive psychologist Bernie Baars postulated that such a processing resource exists in the brain. Its capacity is very small, though, so only a single percept, thought, or memory can be represented at any one time. New information competes with the old and displaces it.

The molecular biologist Jean-Pierre Changeux and the cognitive neuroscientist Stanislas Dehaene, at the Collège de France in Paris, subsequently mapped these ideas onto the architecture of neocortex. The workspace is a network of long-range cortical neurons with reciprocal projections to homologous neurons in other cortical areas, distributed over prefrontal, parieto-temporal, and cingulate associative cortices.

When activity in sensory cortices exceeds a threshold it triggers a global ignition, whereby information enters the global neuronal workspace. This information then becomes available to a host of subsidiary processes such as working memory, language, planning, and voluntary action. The act of globally broadcasting this information is what makes this data conscious. Nothing more and nothing less. Data not broadcast in this manner might still influence behavior but only nonconsciously.

Global workspace theory argues that NCCs occur relatively late (> 350 msec) after stimulus onset and rely on widespread cortical interactions involving frontoparietal networks. It furthermore postulates that attention is necessary for conscious perception, and that working memory is closely linked with global neuronal workspace activity. The model makes experimentally testable predictions that partially overlap with but also diverge strikingly from those of IIT.[16] It is a functionalist account of the mind, not concerned with the causal properties of the underlying system. (That is the Achilles' heel of any purely computational account.)

On this view, consciousness is a consequence of a particular type of algorithm the human brain runs. Conscious states are fully constituted by their functional relationship to relevant sensory inputs, motor outputs, and internal variables, such as those relating to memory, emotion, motivation, vigilance, and so on. The model fully embraces the mythos of our age, to wit:

> Our stance is based on a simple hypothesis: What we call "consciousness" results from specific types of information-processing computations, physically realized by the hardware of the brain.[17]

Since souls and other spooky stuff are ruled out—there is no ghost in the machine—there is no alternative. It doesn't matter whether the hardware is wet neurons or dry-etched transistors. All that matters is the nature of the computations. On this view, appropriately programmed computer simulations of humans will experience their world.

Let me now apply the sharp conceptual scalpel of integrated information theory to dissect the hypothesis that consciousness can be computed. This dissection won't end well for the patient.

13 Why Computers Can't Experience

Barring some catastrophic planetary nightfall, the tech industry will create, within decades, machines with human-level intelligence and behaviors, capable of speech, reasoning, highly coordinated actions in economics, politics and, inevitably, warcraft. The birth of true artificial intelligence will profoundly affect humankind's future, including whether it has any.

Whether you are among those who believe that the arrival of artificial general intelligence signals the dawn of an age of plenty or the sunset of *Homo sapiens*, you will still have to answer the fundamental question: Are these AIs conscious? Does it feel like anything to be them? Or are they just more sophisticated versions of Amazon's Alexa or smartphones—clever machines without any feelings?

Chapters 2 through 4 brought to bear evidence from psychology and neuroscience that intelligence and experience are distinct: being dumb or smart is different from being less or more conscious. In line with this, the neural correlates of consciousness with their center of gravity in the back of cortex are distinct from the correlates of intelligent behavior with their epicenter toward the front (chapter 6). Conceptually, intelligence is about doing while experience is about being, such as being angry or in a state of pure experience. All of this should make us question the unspoken assumption that machine intelligence implies, of necessity, machine consciousness.

With a fundamental theory of consciousness in hand, I will tackle this issue from first principles to demonstrate that intelligence and experience can come apart. Let us apply IIT's postulates to compute how much causal power and integrated information two classes of canonical circuits possess.

The first is a feedforward circuit of the type introduced a couple of pages earlier. Such a neural network, no matter how many processing layers follow each other, is fully reducible. Its integrated information will always be zero. It does not exist intrinsically. The second is a physical realization of a computer, programmed to simulate a network of logic gates. The network it is simulating is irreducible, with non-zero integrated information. Yet, although the computer accurately simulates this irreducible circuit, the computer itself is reducible to its component, without any integrated information, no matter what it simulates.

I will discuss the implications of this fundamental result for whole brain emulation and mind uploading before returning to the distinction between intelligence and consciousness.

Doing the Same but Not Being the Same

What is the integrated information associated with a purely feedforward architecture, in which the output of any one layer of processing elements provides the input to the next processing layer, in a cascade, with no information flowing in the reverse direction? The state of the first layer of the network is determined by outside inputs, say by a camera, and not by the system itself. Likewise, the final processing layer, the system's output, does not affect the rest of the network. That is, from the intrinsic point of view, neither the first nor the last layer of a feedforward network is irreducible. By induction, the same argument holds for the second processing layer and the penultimate output layer and so on. Therefore, considered as a whole, a purely feed-forward network is not integrated. It has no intrinsic causal powers and does not exist for itself, as it is reducible to its individual processing units. A feedforward network never feels like anything, no matter how complex each layer.[1]

Indeed, the intuition that sustained feedback, also called *recurrent* or *reentry processing*, is necessary to experience anything is widespread among neuroscientists.[2] It can now be made precise within the mathematical framework of IIT.

The recurrent network illustrated in figure 13.1 has two input units, six internal processing units, and two outputs. The six core units are richly interconnected with excitatory and inhibitory synapses. Applying IIT's causal calculus to the state shown in figure 13.1 (white signals OFF and gray ON) yields 17 first- and higher-order distinctions (formed by combinations of one or more units within the core). The set of these distinctions forms the maximally irreducible cause-effect structure with a non-zero value of Φ^{max}.

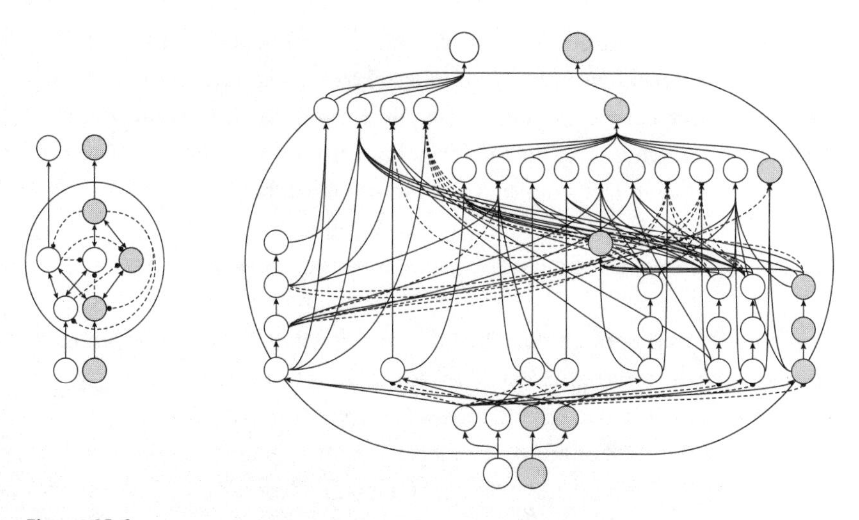

Figure 13.1

Two functional equivalent networks: Two networks that perform the same input–output function can have very different intrinsic causal powers if they differ in their internal wiring. The recurrent net on the left has non-zero integrated information, existing as a Whole, while its fully unfolded twin on the right, with the same input–output mapping, has zero integrated information. It is fully reducible to its 39 individual units. (Redrawn from figure 21 in Oizumi, Albantakis, & Tononi, 2014.)

Now consider the feedforward network in figure 13.1. It too has two inputs and two outputs, but 39 rather than 6 internal processing units and a ton of excitatory and inhibitory connections. This baroque net was laboriously handcrafted to replicate the function of the recurrent network on the left. Both perform the exact same input–output transformations on any input that extends over four time-steps.[3] However, the feedforward circuit has zero Φ^{max} and does not exist as a Whole. Indeed, it is reducible to its 39 atomic elements.

The feedforward circuit is extremely unlikely to have evolved naturally, because all those extra connections and units come at a metabolic cost; furthermore, the circuit is not very robust to damage, as destruction of a single link will often result in failure, while recurrent networks are quite resilient to damage. But it is proof that two networks can perform the same input-output function while having different intrinsic causal powers. The recurrent net is irreducible while its feedforward, functionally identical, version is not. The difference that makes the difference is under the hood, in the system's internal architecture.

Today's success stories in machine learning are feedforward, deep convolutional networks with up to a hundred layers, each one feeding into the next one. They can name dog breeds indistinguishable to most, translate poetry, and imagine visual scenes they have not previously seen.[4] Yet they have no integrated information. They do not exist for themselves.

Digital Computers Have Only Minuscule Intrinsic Existence

Applying IIT to programmable digital computers, we come to an even more startling conclusion that violates strongly held intuitions about functionalism. The maximally irreducible cause-effect power of real physical computers is tiny and independent of the software running on the computer.

Let's follow the work of two brilliant young scholar in Tononi's laboratory, graduate student Graham Findlay and postdoctoral fellow William Marshall,[5] to understand this result. They consider the three-element target network called (PQR) (fig. 13.2) we encountered in chapter 8 (fig. 8.2), but

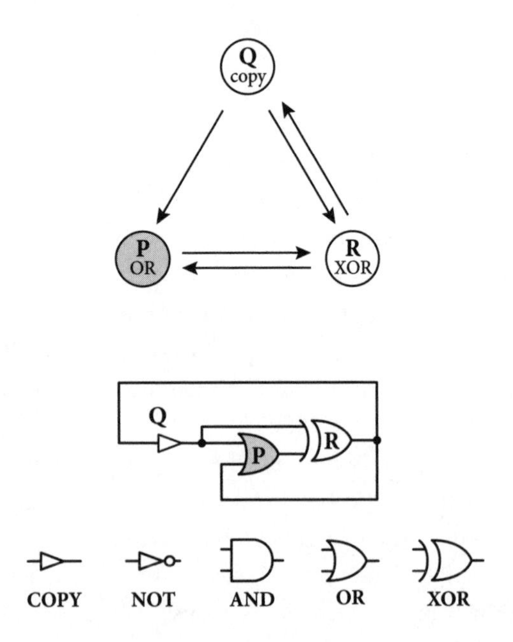

Figure 13.2

An irreducible electronic circuit: The three-node network from figure 8.2 is built from three logic gates. It is a Whole with non-zero integrated information and four distinctions. (From figure 1 in Findlay et al., 2019.)

now implemented with physical circuit elements that instantiate binary gates. Each gate has two input and two output states—high voltage signaling ON or 1 (indicated by gray in figs. 13.2 and 13.3) and low voltage, signaling OFF or 0 (indicated by white). Their internal mechanism implements a logic OR gate, a copy-and-hold unit, and an exclusive OR, or XOR, gate. As this circuit is deterministic, if placed in the state (PQR) = (100), it will shift at the next update into state (001).

Repeating the causal analysis of chapter 8 finds that the system is an irreducible Whole with a non-zero Φ^{max} (fig. 8.2). Its maximally irreducible cause-effect structure is composed of two first-order (Q) and (R), one second-order (PQ), and one third-order mechanism (PQR).

So far so good. Now let us *simulate* this three-element network on the 3-bit programmable computer shown in figure 13.3. Deriving this circuit is a major tour de force. Made out of 66 logic COPY, NOT, AND, OR, and XOR gates, its architecture captures the defining aspects of a classical von Neumann architecture—an arithmetic logic processing unit, a program block, data registers, and a clock that keeps everything nice and tidy in lockstep. The functionality of the simulated circuit is in the program, the eight blocks of four-ring COPY gates, instantiating the eight possible transitions of the three-element circuit (PQR).

It is straightforward but laborious to walk through the operations of this minuscule computer step-by-step as it simulates (PQR).[6] It accurately imitates the triadic circuit until the end of time. That is, the computer in figure 13.3 is functionally equivalent to the target circuit in figure 13.2. Does that imply the same intrinsic causal powers? To answer this, let us apply IIT's causal analysis to the computer.

We're in for a surprise, for the full 66-element circuit is reducible, with zero integrated information. That's because many of the key modules—the clock and the eight rings of four COPY gates—are connected in a feedforward manner to the rest of the circuit. The computer is not a Whole; it has no intrinsic causal powers.

Adding feedback connections within various modules, such as to the clock and the program circuit that retain their functionality, does not change the outcome. The entire computer does not exist for itself.[7] It's not just any old feedback that renders a system irreducible.

We can conceptually break the 66-element computer into all possible smaller circuits and compute Φ^{max} for each chunk. We end up with nine

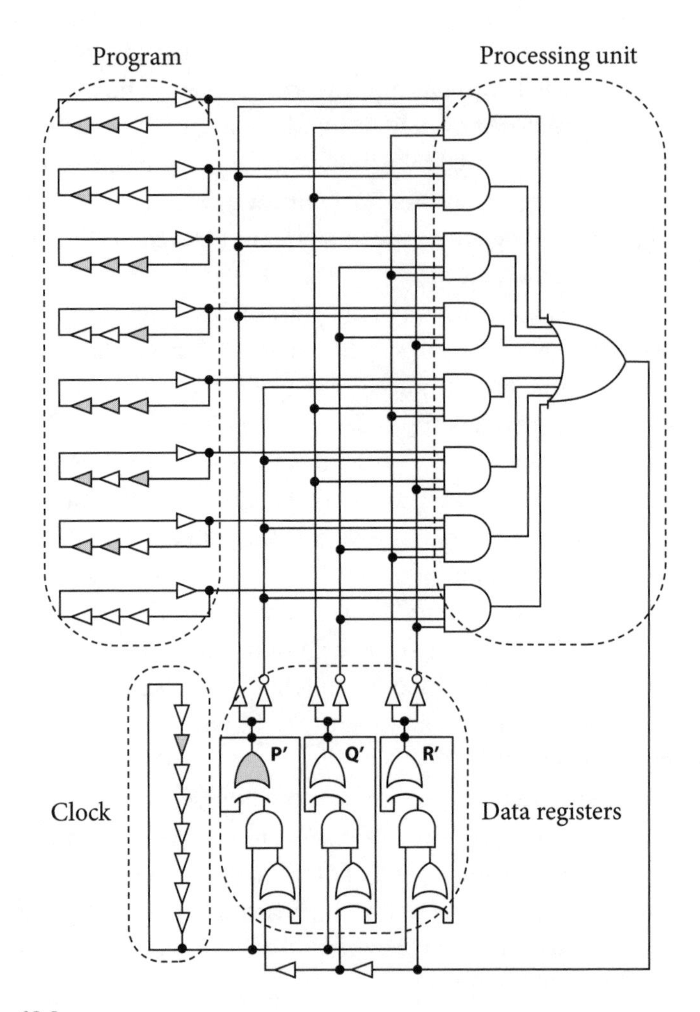

Figure 13.3

A reducible computer simulates an irreducible circuit: A 66-element computer that is functionally equivalent to the triadic circuit of figure 13.2. It can be programmed to simulate any three-element logic circuit. According to IIT, this 3-bit computer does not exist for itself, with zero integrated information, even though it is simulating a circuit with non-zero integrated information. (From figure 2 in Findlay et al., 2019.)

fragments that have intrinsic existence, the clock and the eight rings of four COPY gates. Each of these nine modules constitute a tiny Whole, each with a single first-order mechanism and a minuscule Φ^{max}, a much smaller degree of irreducibility than the physical circuit the computer is simulating.

Why go to all the trouble to build this cumbersome computer that has twenty times more gates and takes eight times longer to run than (PQR)? Because by manipulating the states of the four-ring of COPY gates, the computer can be programmed to simulate any three-gate circuit, not just (PQR)! Despite its apparent simplicity, the computer is universal for this class of circuits.

Findlay and Marshall demonstrate this by analyzing another three-element circuit (XYZ) that implements a computational rule different from (PQR).[8] Analyzed in terms of IIT, this triadic circuit has a maximally irreducible cause-effect structure quite different from that of (PQR), with seven rather than four distinctions. Now, the computer of figure 13.3 is reprogrammed to implement the functionality of (XYZ), which it does perfectly. Yet as with (PQR), the circuit has no intrinsic causal powers, fragmenting into the same nine modules as before.

Let's take stock of the situation. We have two distinct elementary circuits, (PQR) and (XYZ), and a computer that can simulate either one. Per IIT, the triadic circuits have intrinsic causal power and are irreducible, while the much larger computer instantiating these two circuits has no integrated information and can be reduced to smaller modules.

Astute circuit designers will have noticed that the computer, designed to simulate any three element circuit, can be extended to simulate a four element or 4-bit circuit. That requires a total of sixteen ring elements of five COPY gates, feeding into sixteen AND gates that send their output to an OR gate. The clock and data register will have to be extended as well. Indeed, the computer can be upgraded following the same design principles to simulate any finite circuit of n gates, whether it has three or four or 86 billion binary gates.[9] It is Turing complete. Yet no matter its size, intrinsically speaking, it never exists as a Whole, disintegrating into 2^n programming modules and the clock. Each has negligible phenomenal content, independent of the particular circuit it is simulating.

I can't stress enough the complete dissociation between the profoundly impoverished causal structure of the fragmented computer and the potentially rich intrinsic cause-effect structure of the irreducible physical circuit the computer is accurately simulating.

Adding feedback connections here and there doesn't affect this conclusion substantially. The larger the computer, the more obvious its lack of integration owing to its sparse connectivity compared to brains, the lack of internal fan-in and fan-out, its modularity and its serial design.

It could be objected that the computer was not analyzed at the right spatiotemporal level of granularity. After all, by IIT's exclusion postulate, the maximum cause-effect power needs to be evaluated over all possible chunking of space and time and circuit elements. The mathematical machinery of IIT includes a powerful technique to carry out such an analysis, *black-boxing.*[10]

When only the mean behavior of a bunch of variables matters, the variables can be coarse-grained. However, in many cases the exact state of the micro-variables matters a great deal; the particular voltage distribution across millions of cone photoreceptors in your eyes conveys the sense of any one particular visual scene. Averaging over all of them would yield gray. Or consider the electrical charges of the transistor gates in your laptop. Smearing out the total charge over all gates would cause the circuit to cease up and stop working. This is where black-boxing comes in: the low-level functionality is replaced by a black box with specific inputs and output and a particular input–output transformation.

A great example of black-boxing are the three logic gates of figure 13.2. In practice, each one is made out of transistors, resistances, diodes, and other more primitive circuit elements that implement the various logic functions.

Now the heavy lifting of Findlay and Marshall happens.[11] They prove that no black-boxing supports a meaningful maximally irreducible cause-effect structure equivalent to that of the circuit being simulated. Countless arrangements of black-boxing in space, in time (for instance, the eight updates of the clock can be treated as a macro-element in time), and in spacetime are possible, but none does the job: in none does the circuit exist as a Whole.

On the Futility of Mind Uploading

This exposition demonstrates the fallacy of a computational explanation of consciousness. Two systems can be functionally equivalent, they can compute the same input–output function, but they don't share the same intrinsic cause-effect form. The computer of figure 13.3 doesn't exist intrinsically, while the circuit that is being simulated does. That is, they both *do* the same thing, but only one *is* for itself.

Furthermore, the exemplary circuit of figure 13.3 demonstrates that a digital, clocked simulation can completely replicate the function of any target circuit while experiencing next to nothing, no matter what the computer is programmed to do.

Consciousness is not a clever algorithm. Its beating heart is causal power upon itself, not computation. And here's the rub: causal power, the ability to influence oneself or others, cannot be simulated. Not now, nor in the future. It has to be built into the physics of the system.

As an analogy, consider computer code that simulates the field equations of Einstein's theory of general relativity, relating mass to spacetime curvature. Such software can simulate the supermassive black hole at the center of our Milky Way galaxy, Sagittarius A*. This mass exerts such strong gravitational effects on its surroundings that nothing, not even light, can escape it.

But have you ever wondered why the astrophysicists who simulate black holes don't get sucked into their supercomputer? If their models are so faithful to reality, why doesn't spacetime close around the computer doing the modeling, generating a mini black hole that swallows up the computer and everything around it?

Because gravity is not a computation! Gravity has real extrinsic causal power. These powers can be functionally simulated (in terms of an one-to-one mapping between physical properties, such as the metric tensor, the local curvature, and the mass distribution on the one hand and the abstract variables specified at the algorithmic level by the programming language on the other), but that doesn't give these simulations causal power.

Of course, the supercomputer carrying out the relativistic simulations has some mass that will influence spatial curvature by an itsy-bitsy amount. It has a whit of extrinsic causal power; this trifle of causal power will not change when the supercomputer is reprogrammed to run financial spreadsheets, as its mass won't change.

The difference between the real and the simulated is their respective causal powers. That's why it doesn't get wet inside a computer simulating a rain storm. The software can be functionally identical to some aspect of reality, but it won't have the same causal powers as the real thing.[12]

What is true of extrinsic causal power is also true of intrinsic causal power. It is possible to functionally simulate the dynamics of a circuit but its intrinsic cause-effect powers can't be created *ex nihilo*. Yes, the computer, treated as a mechanism, has some minute, intrinsic cause-effect power at the level of the metal, at the level of its transistors, capacitances, and wires. However, the computer does not exist as a Whole but only as tiny fragments. And this is true whether it simulates a black hole or a brain.

This is true even if the simulation would satisfy the most stringent demands of a microfunctionalist. Fast forward a few decades into the future when biophysically and anatomically accurate whole-human-brain emulation technology—of the sort discussed in the previous chapter—can run in real time on computers.[13] Such a simulation will mimic the synaptic and neuronal events that occur when somebody sees a face or hears a voice. Its simulated behavior (for instance, for the sort of experiments outlined in fig. 2.1) will be indistinguishable from those of a human. But as long as the computer simulating this brain resembles in its architecture the von Neumann machine outlined in figure 13.3, it won't see an image; it won't hear a voice inside its circuitry; it won't experience anything. It is nothing but clever programming. Fake consciousness—pretending by imitating people at the biophysical level.

In principle, special-purpose hardware built according to the brain's design principles, so-called *neuromorphic electronic hardware*,[14] could amass sufficient intrinsic cause-effect power to feel like something. That is, if individual logic gates receive inputs from tens of thousands of logic gates and make output connections to tens of thousands of other gates rather than a handful in today's arithmetic logic units[15] and if these massive input and output streams would overlap and feed back onto each other the way neurons do in the brain, then computers' intrinsic cause-effect power could rival that of brains. Such neuromorphic computers could have human-level experience. But that would require a radically different processor layout and a complete conceptual redesign of the entire digital infrastructure of the machine. Once again, it is not by dint of a soul-like substance that brains can experience, but by their causal power upon themselves. Replicate those causal powers and consciousness follows.

Unraveling integrated networks into functionally equivalent feedforward ones illustrates why we can't rely on the famous Turing test to detect machine consciousness. Alan Turing's motivation behind inventing his imitation game was to replace the vague question of "can machines think" with a precise and pragmatic operation, a sort of game show.[16] An agent passes this test when you speak with it for a reasonable amount of time about anything and you can't tell it from a person. The logic is that people "think" when engaged in the give-and-take of a conversation, so if a machine can accomplish the same—if you can converse with it about the weather, the stock market, politics, the local sports team, whether it believes in an afterlife—it should likewise be accorded the same privilege as you, that is, the ability to think. The grandchildren of Alexa and Siri will pass this milestone. However, this does not imply that these programs will feel like anything. Intelligence and consciousness are very different.

Figure 11.2 ranked natural and artificial systems along two axes. The horizontal one represents some operational measure of their intelligence like an IQ test, while their integrated information is plotted along the vertical axis. Figure 13.4 is a variant of this diagram that includes programmable computers. Contemporary software running on conventional digital computers achieves superhuman performance in board games that are traditionally associated with human smarts—IBM's Deep Blue defeated world chess champion Garry Kasparov in 1997 and DeepMind's AlphaGo algorithm defeated a top-seeded Go player, Lee Sedol, in 2016. A supercomputer running a fictitious whole-brain emulation will be as smart as a human. But all of them live along the bottom of the I-C plane—without inner light.

IIT spells doom for the hope that we can transcend brain death by uploading our minds to the cloud. This is based on the idea, popularized by the Princeton neuroscientist Sebastian Seung, of the *connectome*—a record of every one of our trillion synapses and which pair of nerve cells among the brain's 86 billion neurons it connects to. He argues that all of your habits, traits, memories, hopes, and fears have their material residue in the connectome. Seung quips that "you are your connectome." According to computationalism, if your connectome could be uploaded to a future supercomputer that specialized in executing brain simulations, this simulacrum would allow your mind to live on as a pure digital construct, inside the machine.[17]

Leaving aside all scientific and practical objections to this idea, and assuming that we had sufficiently powerful computers (maybe quantum ones) to run this code, uploading would only work if the intrinsic causal powers of these machines matched those of the human brain. Otherwise, you would appear to be living an envious life in utopia, but you wouldn't experience any of it. You would be raptured into digital paradise as a zombie.

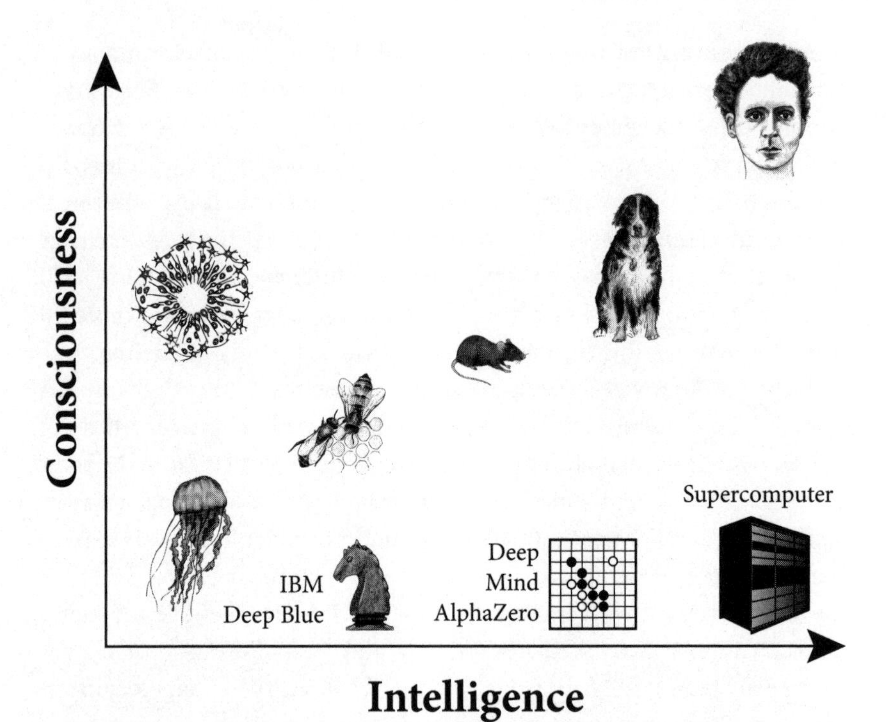

Figure 13.4
Intelligence and consciousness in evolved organisms and engineered artifacts: As species evolve larger nervous systems, their ability to learn and flexibly adapt to new environments, their *intelligence*, increases, as does their capacity to *experience*. Engineered systems deviate in striking ways from this diagonal trend, with increasing digital intelligence but no experience. Bioengineered cerebral organoids may be able to experience something but without being able to do anything (chapter 11).

Will this stop people from opting to upload once the technology is perfected? I doubt it. History provides ample evidence that people are willing to believe some pretty weird things—virginal birth, resurrection from the dead, seventy-two maidens awaiting a suicide bomber in paradise, and so on—in their fervent desire to avoid the end of the reel.

The Expander Graph and Cortical Carpets

This brings me to an objection to integrated information theory by the quantum physicist Scott Aaronson. His argument has given rise to an instructive online debate that accentuates the counterintuitive nature of some of IIT's predictions.[18]

Aaronson estimates Φ^{max} for networks called expander graphs, characterized by being both sparsely yet widely connected.[19] Their integrated information will grow indefinitely as the number of elements in these reticulated lattices increases. This is true even of a regular grid of XOR logic gates. IIT predicts that such a structure will have high Φ^{max}.[20] This implies that two-dimensional arrays of logic gates, easy enough to build using silicon circuit technology, have intrinsic causal powers and will feel like something. This is baffling and defies commonsense intuition. Aaronson therefore concludes that any theory with such a bizarre conclusion must be wrong.

Tononi counters with a three-pronged argument that doubles down and strengthens the theory's claim. Consider a blank featureless wall. From the extrinsic perspective, it is easily described as empty.[21] Yet the intrinsic point of view of an observer perceiving the wall seethes with an immense number of relations. It has many, many locations and neighborhood regions surrounding these. These are positioned relative to other points and regions—to the left or right, above or below. Some regions are nearby, while others are far away. There are triangular interactions, and so on. All such relations are immediately present; they do not have to be inferred. Collectively, they constitute an opulent experience, whether it is seen space, heard space, or felt space. All share a similar phenomenology. The extrinsic poverty of empty space hides vast intrinsic wealth. This abundance must be supported by a physical mechanism that determines this phenomenology through its intrinsic causal powers.

Enter the grid, such as a network of a million integrate-or-fire or logic units arrayed on a 1,000 by 1,000 lattice, somewhat comparable to the output of an eye. Each grid element specifies which of its neighbors were likely ON in the immediate past and which ones will be ON in the immediate future. Collectively, that's one million first-order distinctions. But this is just the beginning, as any two nearby elements sharing inputs and outputs can specify a second-order distinction if their joint cause-effect repertoire cannot be reduced to that of the individual elements. In essence, such a second-order distinctions links the probability of past and future states of the elements' neighbors. By contrast, no second-order distinction is specified by elements without shared inputs and outputs, since their joint cause-effect repertoire is reducible to that of the individual elements. Potentially, there are a million times a million second-order distinctions. Similarly, subsets of three elements, as long as they share input and output, will specify third-order distinctions linking more of their neighbors together. And on and on.

This quickly balloons to staggering numbers of irreducibly higher-order distinctions. The maximally irreducible cause-effect structure associated with such a grid is not so much *representing* space (for to whom is space presented *again*, for that is the meaning of re-presentation?) as creating experienced space from an intrinsic perspective.

Finally, Tononi argues that the neural correlate of consciousness in the human brain resembles a grid-like structure. One of the most robust findings in neuroscience is how visual, auditory, and touch perceptual spaces map in a topographic manner onto visual, auditory, and somatosensory cortices. Most excitatory pyramidal cells and inhibitory interneurons have local axons strongly connected to their immediate neighbors, with the connection probability decreasing with distance.[22] Topographically organized cortical tissue, whether it develops naturally inside the skull or is engineered out of stem cells and grown in dishes, will have high intrinsic causal power. This tissue will feel like something, even if our intuition rebels at the thought that cortical carpets, disconnected from all of their inputs and outputs, can experience anything. But this is precisely what happens to each one of us when we close our eyes, go to sleep, and dream. We create a world that feels as real as the awake one, while devoid of sensory input and unable to move.

Cerebral organoids or grid-like substrates will not be conscious of love or hate but of space; of up, down, close by and far away and other spatial phenomenology distinctions. But unless provided with sophisticated motor outputs, they will be unable to do anything. That is why these grids belong in the upper left-hand corner of the I-C plane.

In the final chapter, I will take stock of the situation and survey who, under time's wide horizon, has intrinsic existence and who does not—not because they can do greater or lesser things but because they have an intrinsic point of view, because they exist for themselves.

14 Is Consciousness Everywhere?

In this closing chapter, I return to the fundamental question I first broached in chapter 2: who else, besides myself, has experiences? Because you are so similar to me, I abduce that you too have subjective, phenomenal states. The same logic applies to other people. Apart from the occasional solitary solipsist this is uncontroversial. But who else has experiences? How widespread is consciousness in the cosmos at large?

I will tackle this question in two ways. The argument by analogy sets out the empirical evidence for the abduction that many species experience the world. This is based on the similarity of their behavior, physiology, anatomy, embryology, and genomics to that of humans, as we are the ultimate arbiter of consciousness.[1] How far consciousness extends its dominion within the tree of life becomes more difficult to abduce as species become more alien to us.

A completely different line of argument takes the principles of integrated information theory to their logical conclusion. Some level of experience can be found in all organisms, including perhaps in *Paramecium* and other single cell life forms. Indeed, according to IIT, experience may not even be restricted to biological entities but might extend to non-evolved physical systems previously assumed to be mindless, a pleasing and parsimonious conclusion about the makeup of the universe.

How Widespread Is Consciousness in the Tree of Life?

The evolutionary relationship among bacteria, fungi, plants, and animals is commonly visualized using the tree of life metaphor (fig. 14.1).[2] All living species, whether fly, mouse, or person, lie somewhere on the periphery of the tree, all equally adapted to their particular ecological niches.

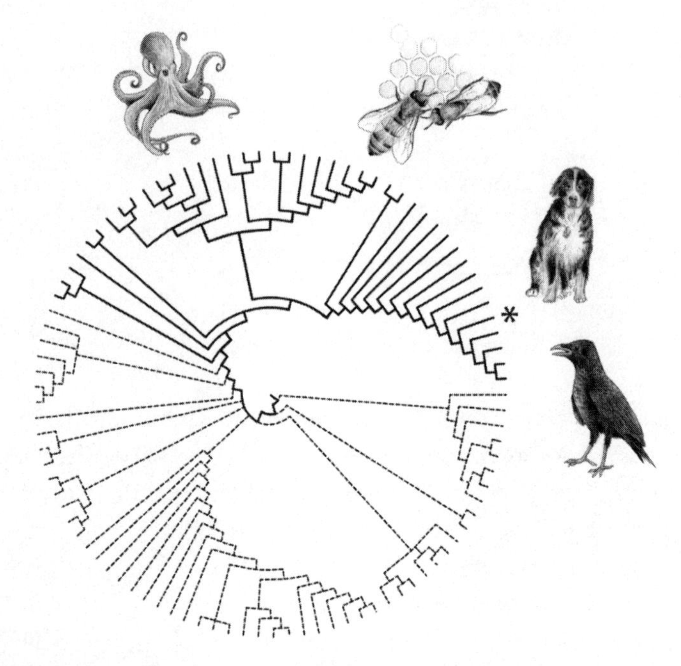

Figure 14.1

The tree of life: Based on the complexity of their behavior and nervous systems, it is likely that it feels like something to be a bird, mammal (marked by *), insect, and cephalopod—represented here by a crow, dog, bee, and octopus. The extent to which consciousness is shared across the entire animal kingdom, let alone across all of life's vast domain, is at present difficult to establish. The last universal common ancestor of all living things is at the center, with time radiating outward.

Every living organism descends in an unbroken lineage from the last universal common ancestor (abbreviated to a charming LUCA) of planetary life. This hypothetical species lived an unfathomable 3.5 billion years ago, smack at the center of the tree-of-life mandala. Evolution explains not only the makeup of our bodies but also the constitution of our minds—for they don't get a special dispensation.

Given the similarities at the behavioral, physiological, anatomical, developmental, and genetic levels between *Homo sapiens* and other mammals, I have no reason to doubt that all of us experience the sounds and sights, the pains and pleasures of life, albeit not necessarily as richly as we do. All of us strive to eat and drink, to procreate, to avoid injury and death; we bask in the sun's warming rays, we seek the company of conspecifics, we fear predators, we sleep, and we dream.

While mammalian consciousness depends on a functioning six-layered neocortex, this does not imply that animals without a neocortex do not feel. Again, the similarities between the structure, dynamics, and genetic specification of nervous systems of all tetrapods—mammals, amphibians, birds (in particular ravens, crows, magpies, parrots), and reptiles—allows me to abduce that they too experience the world. A similar inference can be made for other creatures with a backbone, such as fish.[3]

But why be a vertebrate chauvinist? The tree of life is populated by a throng of invertebrates that move about, sense their environment, learn from prior experience, display all the trappings of emotions, communicate with others—insects, crabs, worms, octopuses, and on and on. We might balk at the idea that tiny buzzing flies or diaphanous pulsating jellyfish, so foreign in form, have experiences.

Yet honey bees can recognize faces, communicate the location and quality of food sources to their sisters via the waggle dance, and navigate complex mazes with the help of cues they store in short-term memory. A scent blown into a hive can trigger a return to the place where the bees previously encountered this odor, a type of associative memory. Bees have collective decision-making skills that, in their efficiency, put any academic faculty committee to shame. This "wisdom of the crowd" phenomenon has been studied during swarming, when a queen and thousands of her workers split off from the main colony and chooses a new hive that must satisfy multiple demands crucial to group survival (think of that when you go house hunting). Bumble bees can even learn to use a tool after watching other bees use them.[4]

Charles Darwin, in a book published in 1881, wanted "to learn how far the worms acted consciously and how much mental power they displayed."[5] Studying their feeding behaviors, Darwin concluded that there was no absolute threshold between complex and simple animals that assigned higher mental powers to one but not to the other. No one has discovered a Rubicon that separates sentient from nonsentient creatures.

Of course, the richness and diversity of animal consciousness will diminish as their nervous system becomes simpler and more primitive, eventually turning into a loosely organized neural net. As the pace of the underlying assemblies becomes more sluggish, the dynamics of the organisms' experiences will slow down as well.

Does experience even require a nervous system? We don't know. It has been asserted that trees, members of the kingdom of plants, can communicate with each other in unexpected ways, and that they adapt and learn.[6] Of course, all of that can happen without experience. So I would say the evidence is intriguing but very preliminary. As we step down the ladder of complexity rung by rung, how far down do we go before there is not even an inkling of awareness? Again, we don't know. We have reached the limits of abduction based on similarity with the only subject we have direct acquaintance with—ourselves.[7]

Consciousness in the Universe

IIT offers a different chain of reasoning. The theory precisely answers the question of who can have an experience: anything with a non-zero maximum of integrated information;[8] anything that has intrinsic causal powers is a Whole. What this Whole feels, its experience, is given by its maximally irreducible cause-effect structure (fig. 8.1). How much it exists is given by its integrated information.

The theory doesn't stipulate that Φ^{max} has to exceed 42 or any other magical threshold for experience to switch on. Anything with Φ^{max} greater than zero exists for itself, has an inner view, and has some degree of irreducibility. And that means there are a lot of Wholes out there.

Certainly, this includes people and other mammals with neocortex, which we clinically know to be the substrate of experience. Fish, birds, reptiles, and amphibians possess a telencephalon that is evolutionarily related to mammalian cortex. Given the attendant circuit complexity, the intrinsic causal power of the telencephalon is likely to be high.

When considering the neural architecture of creatures very different from us, such as the honeybee, we are confronted by vast and untamed neuronal complexity—about one million neurons within a volume the size of a grain of quinoa, a circuit density ten times higher than that of our neocortex of which we are so proud. And unlike our cerebellum, the bee's mushroom-shaped body is heavily recurrently connected. It is likely that this little brain forms a maximally irreducible cause-effect structure.

Integrated information is not about input–output processing, function or cognition, but about intrinsic cause-effect power. Having liberated itself from the myth that consciousness is intimately related to intelligence (chapters 4, 11, and 13), the theory is free to discard the shackles of nervous systems and to locate intrinsic causal power in mechanisms that do not compute in any conventional sense.

A case in point is that of single-cell organisms, such as *Paramecium*, the *animalcules* discovered by the early microscopists in the late seventeenth century. Protozoa propel themselves through water by whiplash movements of tiny hairs, avoid obstacles, detect food, and display adaptive responses. Because of their minuscule size and strange habitats, we don't think of them as sentient. But they challenge our presuppositions. One of the early students of such microorganisms, H. S. Jennings, expressed this well:

> The writer is thoroughly convinced, after long study of the behavior of this organism, that if Amoeba were a large animal, so as to come within the everyday experience of human beings, its behavior would at once call forth the attribution to it of states of pleasure and pain, of hunger, desire, and the like, on precisely the same basis as we attribute these things to the dog.[9]

Among the best-studied of all organisms are the even smaller *Escherichia coli*, bacteria that can cause food-poisoning. Their rod-shaped bodies, about the size of a synapse, house several million proteins inside their protective cell wall. No one has modeled in full such vast complexity. Given this byzantine intricacy, the causal power of a bacterium upon itself is unlikely to be zero.[10] Per IIT, it is likely that it feels like something to be a bacterium. It won't be upset about its pear-shaped body; no one will ever study the psychology of a microorganism. But there will be a tiny glow of experience. This glow will disappear once the bacterium dissolves into its constituent organelles.

Let us travel down further in scale, transitioning from biology to the simpler worlds of chemistry and physics, and compute the intrinsic causal power of a protein molecule, an atomic nucleus or even a single proton. Per the standard model of physics, protons and neutrons are made out of three quarks with fractional electrical charge. Quarks are never observed by themselves. It is therefore possible that atoms constitute an irreducible Whole, a modicum of "enminded" matter. How does it feel to be a single atom compared to the roughly 10^{26} atoms making up a human brain? Given that its integrated information is presumably barely above zero, just a minute bagatelle, a this-rather-than-not-this?[11]

To wrap your mind around this possibility that violates Western cultural sensibilities, consider an instructive analogy. The average temperature of the universe is determined by the afterglow left over from the Big Bang, the cosmic microwave background radiation. It evenly pervades space at an effective temperature of 2.73° above absolute zero. This is utterly frigid, hundreds of degrees colder than any temperature terrestrial organisms can survive. But the fact that the temperature is non-zero implies a corresponding tiny amount of heat in deep space. Likewise, a non-zero Φ^{max} implies a corresponding tiny amount of experience.

To the extent that I'm discussing the mental with respect to single-cell organisms let alone atoms, I have entered the realm of pure speculation, something I have been trained all my life as a scientist to avoid. Yet three considerations prompt me to cast caution to the wind.

First, these ideas are straightforward extensions of IIT—constructed to explain human-level consciousness—to vastly different aspects of physical reality. This is one of the hallmarks of a powerful scientific theory—predicting phenomena by extrapolating to conditions far from the theory's original remit. There are many precedents—that the passage of time depends on how fast you travel, that spacetime can break down at singularities known as black holes, that people, butterflies, vegetables, and the bacteria in your gut use the same mechanism to store and copy their genetic information, and so on.

Second, I admire the elegance and beauty of this prediction.[12] The mental does not appear abruptly out of the physical. As Leibniz expressed it, *natura non facit saltus*, or nature does not make sudden leaps (Leibniz was, after all, the co-inventor of infinitesimal calculus). The absence of discontinuities is also a bedrock element of Darwinian thought.

Intrinsic causal power does away with the challenge of how mind emerges from matter. IIT stipulates that it is there all along.

Third, IIT's prediction that the mental is much more widespread than traditionally assumed resonates with an ancient school of thought: *panpsychism*.

Many but Not All Things Are Enminded

Common to panpsychism in its various guises is the belief that soul (*psyche*) is in everything (*pan*), or is ubiquitous; not only in animals and plants but all the way down to the ultimate constituents of matter—atoms, fields, strings, or whatever. Panpsychism assumes that any physical mechanism either is conscious, is made out of conscious parts, or forms part of a greater conscious whole.

Some of the brightest minds in the West took the position that matter and soul are one substance. This includes the pre-Socratic philosophers of ancient Greece, Thales, and Anaxagoras. Plato espoused such ideas, as did the Renaissance cosmologist Giordano Bruno (burned at the stake in 1600), Arthur Schopenhauer, and the twentieth-century paleontologist and Jesuit Teilhard de Chardin (whose books, defending evolutionary views on consciousness, were banned by his church until his death).

Particularly striking are the many scientists and mathematicians with well-articulated panpsychist views. Foremost, of course, is Leibniz. But we can also include the three scientists who pioneered psychology and psychophysics—Gustav Fechner, Wilhelm Wundt, and William James—and the astronomer and mathematicians—Arthur Eddington, Alfred North Whitehead, and Bertrand Russell. With the modern devaluation of metaphysics and the rise of analytic philosophy, the last century evicted the mental entirely, not only from most university departments but also from the universe at large. But this denial of consciousness is now being viewed as the "Great Silliness," and panpsychism is undergoing a revival within the academe.[13]

Debates concerning what exists are organized around two poles: materialism and idealism. Materialism, and its modern version known as *physicalism*, has profited immensely from Galileo Galilei's pragmatic stance of removing mind from the objects it studies in order to describe and quantify nature from the perspective of an outside observer. It has done so at the cost of ignoring the central aspect of reality—experience. Erwin Schrödinger, one of the founders of quantum mechanics, after whom its most famous equation is named, expressed this clearly:

> The strange fact [is] that on the one hand all our knowledge about the world around us, both that gained in everyday life and that revealed by the most carefully planned and painstaking laboratory experiments, rests entirely on immediate sense perception, while on the other hand this knowledge fails to reveal the relations of the sense perceptions to the outside world, so that in the picture or model we form of the outside world, guided by our scientific discoveries, all sensual qualities are absent.[14]

Idealism, on the other hand, has nothing productive to say about the physical world, as it is held to be a figment of the mind. Cartesian dualism accepts both in a strained marriage in which the two partners live out their lives in parallel, without speaking to each other (this is the *interaction* problem: how does matter interact with the ephemeral mind?).[15] Behaving like a thwarted lover, analytic, logical-positivist philosophy denies the legitimacy and, in its more extreme version, even the very existence of one partner in the mental-physical relationship. It does so to obfuscate its inability to deal with the mental.

Panpsychism is unitary. There is only one substance, not two. This elegantly eliminates the need to explain how the mental emerges out of the physical and vice versa. Both coexist.

But panpsychism's beauty is barren. Besides claiming that everything has both intrinsic and extrinsic aspects, it has nothing constructive to say about the relationship between the two. Where is the experiential difference between one lone atom zipping around in interstellar space, the hundred trillion trillion making up a human brain, and the uncountable atoms making up a sandy beach? Panpsychism is silent on such questions.

IIT shares many insights with panpsychism, starting with the fundamental premise that consciousness is an intrinsic, fundamental aspect of reality. Both approaches argue that consciousness is present across the animal kingdom to varying degrees.

All else being equal, integrated information, and with it the richness of experience, increases as the complexity of the associated nervous system grows (figs. 11.2 and 13.4), although sheer number of neurons is not a guarantee, as shown by the cerebellum. Consciousness waxes and wanes diurnally with alertness and sleep. It changes across the lifespan—becoming richer as we grow from a fetus into a teenager and mature into an adult with a fully developed cortex. It increases when we become familiar with romantic and sexual relationships, with alcohol and drugs, and when we acquire appreciation for games, sports, novels, and art; and it will slowly disintegrate as our aging brains wear out.

Most importantly, though, IIT is a scientific theory, unlike panpsychism. IIT predicts the relationship between neural circuits and the quantity and quality of experience, how to build an instrument to detect experience (chapter 9), pure experience and how to enlarge consciousness by brain-bridging (chapter 10), why certain parts of the brain have it and others not (the posterior cortex versus the cerebellum), why brains with human-level consciousness evolved (chapter 11), and why conventional computers have only a tiny bit of it (chapter 13).

When lecturing about these matters, I often get the you've-got-to-be-kidding-stare. This passes once I explain how neither panpsychism nor IIT claim that elementary particles have thoughts or other cognitive processes. Panpsychism does, however, have an Achilles' heel—the *combination* problem, a problem that IIT has squarely solved.

On the Impossibility of Group Mind, or Why Your Neurons Are Not Conscious

James gave a memorable example of the combination problem in the foundational text of American psychology, *The Principles of Psychology* (1890):

> Take a sentence of a dozen words, and take twelve men and tell to each one word. Then stand the men in a row or jam them in a bunch, and let each think of his word as intently as he will; nowhere will there be a consciousness of the whole sentence.[16]

Experiences do not aggregate into larger, superordinate experiences. Closely interacting lovers, dancers, athletes, soldiers, and so on do not give rise to a group mind, with experiences above and beyond those of the individuals making up the group. John Searle wrote:

> Consciousness cannot spread over the universe like a thin veneer of jam; there has to be a point where my consciousness ends and yours begins.[17]

Panpsychism has not provided a satisfactory answer as to why this should be so. But IIT does. As discussed extensively in chapter 10 with respect to the split-brain experiments (fig. 10.2), IIT postulates that only maxima of integrated information exist. This is a consequence of the exclusion axiom—any conscious experience is definite, with borders. Certain aspects of experience are in, while a vast universe of possible feelings are out.

Consider figure 14.2, in which I'm looking at Ruby and have a particular visual experience (fig. 1.1), a maximally irreducible cause-effect structure. It is constituted by the underlying physical substrate, the Whole, here a particular neural correlate of consciousness within the hot zone in my posterior cortex. But the experience is not identical to the Whole. My experience is not my brain.

This Whole has definite borders; a particular neuron is either part of it or not. The latter is true even if this neuron provides some synaptic input to the Whole. What defines the Whole is a maximum of integrated information,

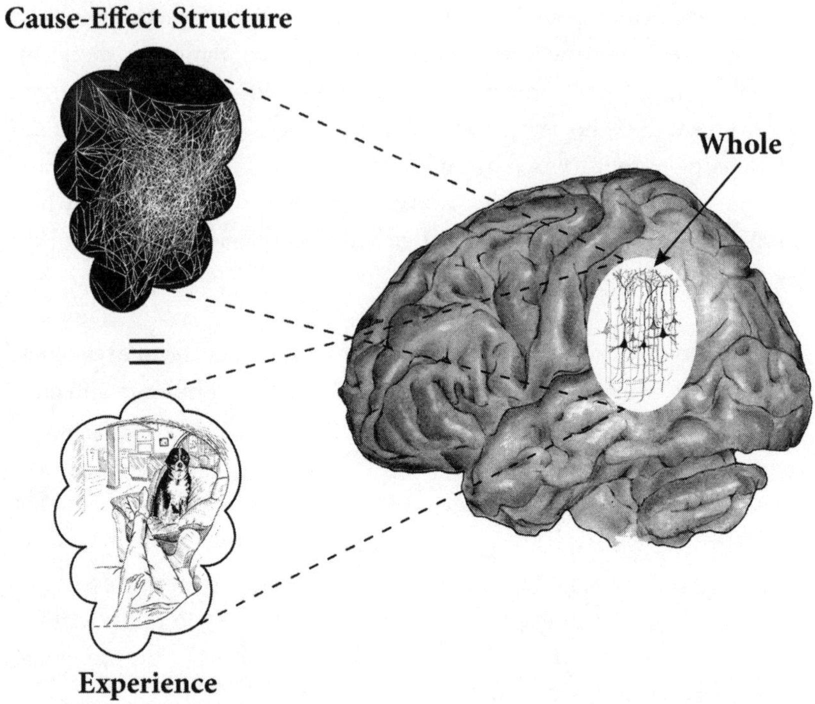

Cause-Effect Structure

Whole

≡

Experience

Figure 14.2

The mind–body problem resolved? Integrated information theory posits that any one conscious experience, here that of looking at a Bernese mountain dog, is identical to a maximally irreducible cause-effect structure. Its physical substrate, its Whole, is the operationally defined neural correlate of consciousness. The experience is formed by the Whole but is not identical to it.

with the maximum being evaluated over all spatiotemporal scales and levels of granularities, such as molecules, proteins, subcellular organelles, single neurons, large ensembles of them, the environment the brain interacts with, and so on.

It is the irreducible Whole that forms my conscious experience, not the underlying neurons.[18] So not only is my experience not my brain, but most certainly it is not my individual neurons. While a handful of cultured neurons in a dish may have an itsy-bitsy amount of experience, forming a mini-mind, the hundreds of millions neurons making up my posterior cortex do not embody a collection of millions of mini-minds. There is only one mind, my mind, constituted by the Whole in my brain.

Other Wholes may exist in my brain, or my body, as long as they don't share elements with the posterior hot zone Whole. Thus, it may feel like something to be my liver, but given the very limited interactions among liver cells, I doubt it feels like a lot.

Likewise, while isolated bacteria might well have a whit of integrated information, the trillion bacteria living happily in my gut would only have a single mind of their own if the associated Φ^{max} of the microbiome considered as a whole is larger than the Φ^{max} of individual bacteria. This is not easy to decide *a priori*, but depends on the strength of the various interactions.

The exclusion principle also explains why consciousness ceases during slow sleep. At this time, delta waves dominate the EEG (fig. 5.1) and cortical neurons have regular hyperpolarized down-states during which they are silent, interspersed by active up-states when neurons are more depolarized. These on- and off-periods are regionally coordinated. As a consequence, the cortical Whole breaks down, shattering into small cliques of interacting neurons. Each one probably has only a whit of integrated information. Effectively, "my" consciousness vanishes in deep sleep, replaced by myriad of tiny Wholes, none of which is remembered upon awakening.[19]

The exclusion postulate also dictates whether or not an aggregate of conscious entities—ants in a colony, cells making up a tree, bees in a hive, starlings in a murmurating flock, an octopus with its eight semiautonomous arms, or the hundreds of Chinese dancers and musicians during the choreographed opening ceremony of the 2008 Olympic games in Beijing—exist as conscious entities. A herd of buffalo during a stampede or a crowd can act as if it had "one mind," but this remains a mere figure of speech unless there is a phenomenal entity that feels like something above and beyond the experiences of the individuals making up the group. Per IIT, this would require the extinction of the individual Wholes, as the integrated information for each of them is less than the Φ^{max} of the Whole. Everybody in the crowd would give up his or her individual consciousness to the mind of the group, like being assimilated into the hive mind of the Borg in the *Star Trek* universe.

IIT's exclusion postulate does not permit the simultaneous existence of both individual and group mind. Thus, the *Anima Mundi* or world soul is ruled out, as it requires that the mind of all sentient beings be extinguished in favor of the all-encompassing soul. Likewise, it does not feel like anything to be the three hundred million citizens of the United States of America. As an entity, the United States has considerable extrinsic causal powers, such as the power to execute its citizens or start a war. But the country does not have maximally irreducible intrinsic cause-effect power. Countries, corporations, and other group agents exist as powerful military, economic, financial, legal, and cultural entities. They are aggregates but not Wholes. They have no phenomenal reality and no intrinsic causal power.[20]

Thus, per IIT, single cells may have some intrinsic existence, but this does not necessarily hold for the microbiome or trees. Animals and people exist for themselves, but herds and crowds do not. Maybe even atoms exist for themselves, but certainly not spoons, chairs, dunes, or the universe at large.

IIT posits two sides to every Whole: an exterior aspect, known to the world and interacting with other objects, including other Wholes; and an interior aspect, what it feels like, its experience. It is a solitary existence, with no direct windows into the interior of other Wholes. Two or more Wholes can fuse to give rise to a larger Whole but at the cost of losing their previous identity.

Finally, panpsychism has nothing intelligible to say about consciousness in machines. But IIT does. Conventional digital computers, built out of circuit components with sparse connectivity and little overlap among their inputs and their outputs, do not constitute a Whole (chapter 13). Computers have only a tiny amount of highly fragmented intrinsic cause-effect power, no matter what software they are executing and no matter their computational power. Androids, if their physical circuitry is anything like today's CPUs, cannot dream of electric sheep. It is, of course, possible to build computing machinery that closely mimics neuronal architectures. Such neuromorphic engineering artifacts could have lots of integrated information. But we are far from those.

IIT can be thought of as an extension of physics to the central fact of our lives—consciousness.[21] Textbook physics deals with the interaction of objects with each other, dictated by extrinsic causal powers. My and your experiences are the way brains with irreducible intrinsic causal powers feel like from the inside.

IIT offers a principled, coherent, testable, and elegant account of the relationship between these two seemingly disparate domains of existence—the physical and the mental—grounded in extrinsic and intrinsic causal powers. Causal power of two different kinds is the only sort of stuff needed to explain everything in the universe.[22] These powers constitute ultimate reality.

Further experimental work will be essential to validate, modify, or perhaps even reject these views. If history is any guide, future discoveries in laboratories and clinics, or perhaps off-planet, will surprise us.

We have come to the end of our voyage. Illuminated by the light of our pole star—consciousness—the universe reveals itself to be an orderly place. It is far more enminded than modernity, blinded by its technological supremacy over the natural world, takes it to be. It is a view more in line with earlier traditions that respected and feared the natural world.

Experience is in unexpected places, including in all animals, large and small, and perhaps even in brute matter itself. But consciousness is not in digital computers running software, even when they speak in tongues. Ever-more powerful machines will trade in fake consciousness, which will, perhaps, fool most. But precisely because of the looming confrontation between natural, evolved and artificial, engineered intelligence, it is absolutely essential to assert the central role of feeling to a lived life.

Coda: Why This Matters

My lifelong quest is to grasp the true nature of being. I have struggled to comprehend how consciousness, so long estranged from science, can fit within a rational, consistent, and empirically testable worldview informed by physics and biology. I have come to a measure of understanding of this question, within the limitations unique to me and to my kind.

I now know that I live in a universe in which the inner light of experience is far more widespread than assumed within the standard Western canon. This inner light shines in humans and in the denizens of the animal kingdom, brighter or dimmer, in proportion to the complexity of their nervous system. Integrated information theory predicts the possibility that all cellular life feels like something. The mental and the physical are closely linked, two aspects of one underlying reality.

These are insights of relevance for their philosophical, scientific, and aesthetic value. However, I'm not just a scientist; I also strive to live an ethical life. What moral consequences flow from this abstract understanding? I want to end this book by moving from the descriptive to the prescriptive and proscriptive, to how we should think about good and bad and to a call to action.[1]

Most importantly, we must abandon the idea that humans are at the center of the ethical universe and that we bestow value on the rest of the natural world only insofar as it suits humanity's ends, a belief that is such a large part of Western culture and tradition.

We are evolved creatures, one leaf among millions on the tree of life. Yes, we are endowed with powerful cognitive abilities, in particular, language, symbolic thinking, and a strong sense of "I." These fuel accomplishments that our relatives on the tree of life are incapable of—science, *Der Ring des Nibelungen*, universal human rights, but also the Holocaust and global warming. Though we are at the threshold of a trans- and posthumanist world in which our bodies will increasingly be intertwined with machines, we remain within the gravitational pull of our biology.[2]

We must cure humanity's narcissism and our deep-seated belief that animals and plants exist solely for our pleasure and benefit. We need to embrace the principle that the moral status of any subject, of any Whole, is rooted in their consciousness, not solely in their humanness.[3] There are three justifications for being admitted to the privileged rank of subjects. I call them the *sentience, experiential,* and the *cognitive* criteria.

Emotionally, we most easily resonate with *sentience* in others. We all have a strong instinctual reaction when we see a child or a dog being abused. We feel something of its pain; we empathize with it. Hence the moral intuition that any subject capable of suffering ceases to be a means to an end and becomes an end in itself. Any creature that has the potential to feel distress has some minimal moral standing—foremost the desire to be and the desire not to suffer.

For a subject to suffer necessarily implies that the subject has experiences. But the opposite is not necessarily true. We can imagine Wholes that feel like something, such as brain organoids and other bio-engineered structures, without painful or adverse experiences that they seek to avoid. Expressed differently, the set of subjects that can suffer is a subset of all subjects. Most higher animals, though, are condemned to experience the pain of living, as monitoring threats to the integrity of one's body and deviation from homeostasis are prime directives for survival.

European modernity created animal rights laws because of society's growing realization that pets and livestock are in some ways similar to people. All are, to a greater or lesser extent, capable of suffering pain and enjoying life, of seeing, smelling, and hearing the world. All have an intrinsic value. Yet while our animal companions, and a handful of charismatic megafauna such as great apes, whales, lions, wolves, elephants, and bald eagles, have public defenders and laws to protect them, this is not the case for reptiles, amphibians, fish, or invertebrates, such as squids, octopuses, or lobsters.

Fish have few advocates for their welfare as they can't shout or scream, are cold-blooded, and look so different from us. We don't even accord them the right to a swift death—fishermen think nothing of impaling fish as live, wiggling bait or of dumping thousands of them out of the net that caught them onto a trawler, leaving them to suffocate in agony. And yes, all physiological, hormonal, and behavioral evidence implies that fish respond to painful stimuli in the same manner as we do. In this ghastly and unthinking way, we kill about a trillion fish each year; a thousand billion sentient creatures.[4] If the arc of the moral universe bends toward justice, humanity will have to account for this routine atrocity in which all of us are implicated.

The *experiential* justification for the idea that species other than humans ought to be considered intrinsically valuable subjects is that any entity with an inner point of view is precious. Phenomenal experience is irreplaceable in a universe devoid of extrinsic purpose. It is the only thing that really matters. For if something does not feel, it does not exist as a subject. A corpse and a zombie both exist, but not for themselves, only for others.

A third justification for this grounding is *cognitive ability*, having beliefs and desires, a sense of self, a sense of the future, of being able to imagine counterfactuals ("If I hadn't lost my legs, I could still climb"), the potential to be creative. To the extent that animals have such advanced cognitive abilities, they have rights.

However, I am against justifying moral standing purely on the basis of cognitive skills. First, not all humans possess such abilities—think of babies, anencephalic children, or patients in a vegetative state or in late-stage dementia. Should they have lesser standing and fewer rights than able adults? Second, only a few species would join *Homo sapiens* in this elite cognitive club. I doubt that my dog has much of a sense of the upcoming weekend. Finally, if we link moral rights to certain functional abilities, such as imagination or intelligence, than we will sooner or later have to admit software running on digital computers into the club. What happens if they become capable of cognitive feats people cannot match? They would then leave us morally in the dust, even though, per IIT, they would not feel anything.

I'm not a carbon chauvinist arguing for the inherent superiority of any and all organic lifeforms over engineered silicon variants. Computers based on neuromorphic architectures could, at least in principle, possess intrinsic causal powers that rival those of large brains with the attendant legal and ethical privileges.

Of these three justifications, sentience is in my view the most powerful one for why creatures that can suffer deserve special status. Because I can imagine myself being left alone, starved, or beaten, I feel sympathy or *empathy* with the other. Empathy is, of course, a conscious experience inconceivable to nonconscious entities, such as software.[5]

Just because two species can suffer does not imply that they suffer to the same degree or with the same intensity. A large gap exists between the complexity of the brain of a fly and a human and their respective conscious experiences. The moral privileges we accord to species should reflect this reality. They are not all situated on the same moral rung.

IIT can rank species according to the quantity of their consciousness, a modern version of the "Great Chain of Being." The ancients called this strict hierarchy *scala naturae*. Grounded in base earth, it ascended through minerals, plants, animals, to humans (from commoners to the king), followed by renegades and true angels. The apotheosis was, of course, the supreme and most perfect being, God.

IIT's ladder of being is defined by the integrated information of any species, Φ^{max}, its irreducibility or how much it exists for itself.[6] This ladder has as its lowest rung aggregates without experience, followed by jellyfish, bees, mice, dogs, and people (see figs. 11.2 or 13.4). It has no natural upper limit. Who knows what we will discover under alien skies or what artifacts we will build in the centuries to come? As Lord Dunsany wrote, "A man is a very small thing, and the night is very large and full of wonders."

I understand the squeamishness that such a ranking provokes. However, we must take the graded nature of the capacity to experience into account if we are to balance the interests of all creatures against each other.

The principle of sentience is a clarion call to act in both the private and the public spheres. The most immediate response to these insights is to stop eating animals, or at least those with high integrated information. Becoming a vegetarian eliminates an enormous amount of suffering pertaining to factory farming from the world.

The writer David Foster Wallace, in his 2004 essay "Consider the Lobster" for *Gourmet* magazine, makes a case that boiling lobsters alive is hideous to them. He rhetorically asks:

> Is it not possible that future generations will regard our own present agribusiness and eating practices in much the same way we now view Nero's entertainments or Aztec sacrifices?

Turning away from flesh also reduces the massive adverse environmental and ecological footprint of industrial-scale animal husbandry; as an added benefit, avoiding flesh improves our physical and mental well-being. A more radical response is to reject any animal products for food or clothing—to become a vegan. This is more difficult.

As our knowledge of suffering in the world increases, we must push for laws that enlarge the circle of creatures that are afforded legal protection. We need a new Decalogue, a new anthropology, and a new moral code as forcefully advocated by the Australian philosopher and ethicist Peter Singer.[7]

There are hopeful signs for a change. Most countries do have anticruelty laws for companion species. Organizations such as the "Great Ape Project" or the "Nonhuman Rights Project" seek to extend some minimal legal rights to big-brain animals—great apes, elephants, dolphins, and whales. Most law schools now offer classes on animal rights. So far, though, no species outside *Homo sapiens* has the standing of *legal persons* with all the associated privileges. Corporations do, but no domesticated or wild animals. They are property under the law.

The New Testament reminds us in Matthew 25: "Truly I tell you, whatever you did for one of the least of these brothers and sisters of mine, you did for me." The Czech writer Milan Kundera in his novel *The Unbearable Lightness of Being* generalizes this moral stipulation to all creatures:

> True human goodness, in all its purity and freedom, can come to the fore only when its recipient has no power. Mankind's true moral test, its fundamental test ... consists of its attitude toward those who are at its mercy: animals.

The Buddhist attitude to sentience mirrors this view. We should treat all animals as being conscious, as feeling what-it-is-like-to-be. This must inform how we behave toward our fellow travelers through this cosmos, who are defenseless before our fences, cages, blades, and bullets and our relentless drive for *Lebensraum*. One day, humanity may well be judged for how we treated our relatives on the tree of life. We should extend a universal ethical stance to all creatures, whether they speak, cry, bark, whine, howl, bellow, chirp, shriek, buzz, or are mute. For all experience life, book-ended between two eternities.

Notes

Chapter 1

1. Koch (2004), p. 10.

2. Descartes first makes this statement in *Discourse on the Method* (1637), as *Je pense, donc je suis*, later on translated as *cogito, ergo sum*. In the second mediation of his *Meditations on First Philosophy* (1641), he expands on this idea: "But what then am I? A thing which thinks. What is a thing which thinks? It is a thing which doubts, understands, [conceives], affirms, denies, wills, refuses, which also imagines and senses." This makes it clear that "*Je pense*" refers to more than just thinking but the entire gamut of mental activities associated with consciousness. There are few truly original ideas, and this one is no exception. Thus, Thomas Aquinas wrote in the middle of the thirteenth century in *Disputed Questions on Truth*: "No one can think that he does not exist and assent: in the very fact that he thinks anything, he perceives himself to exist." See also the following footnote.

3. Saint Augustine of Hippo (*City of God*, Book 11, chapter 26) writes: "But, without any delusive representation of images or phantasms, I am most certain that I am, and that I know and delight in this. In respect of these truths, I am not at all afraid of the arguments of the Academicians, who say, What if you are deceived? For if I am deceived, I am (*si enim fallor, sum*). For he who is not, cannot be deceived; and if I am deceived, by this same token I am. And since I am if I am deceived, how am I deceived in believing that I am? For it is certain that I am if I am deceived." Continuing this historical-archeological sleuthing, consider Aristotle in his *Nicomachean Ethics*: "We sense that we sense and we understand that we understand, and because we sense this, we understand that we exist." I could also cite Parmenides, at which point I've reached the bedrock of recorded Western philosophical thought.

4. Patricia Churchland (1983, 1986) and Paul Churchland (1984). Rey (1983, 1991) and Irvine (2013) likewise articulate such eliminativist views.

5. Dennett argues in his landmark *Consciousness Explained* (1991) and in chapter 14 of his *From Bacteria to Bach and Back* (2017) that people are terribly confused about their

experiences. What they really mean when they speak about consciousness is that they have certain beliefs about mental states; each one has distinct functional properties with distinct behaviors and affordances. Once these outcomes are explained, there is nothing left to account for. There is nothing intrinsic about pain or redness; consciousness is all in the doing. Dennett's frequent invocation of the word "illusion" (*nomen est omen*) and his take that consciousness is real but without any intrinsic properties is not only incoherent but a position so incongruous with my lived experience that I can't properly describe it. He charges people like me, who insist on the intrinsic, authentic nature of consciousness with "hysterical realism." In my opinion, a more accurate title of Dennett's book would be *Consciousness Explained Away*. His writing is characterized by the skillful use of colorful metaphors, analogies, and historical allusions. These literary devices are memorable, vivid, and effective at capturing the reader's imagination. Yet they are difficult to relate to underlying mechanisms. Francis Crick warned me of scholars who write too well—here he was specifically referring to Sigmund Freud (who was nominated for the Nobel Prize in Literature) and Dennett. Fallon (2019a) offers a sympathetic interpretation of Dennett's position on consciousness. Note that there are many distinct strands of thought that deny deeply held intuitions about consciousness, including eliminative materialism, fictionalism, and instrumentalism.

6. Searle (1992), p. 3.

7. Strawson (1994), p. 53. In 1929, the cosmologist and philosopher Alfred Whitehead thundered against those "who arrogated to themselves the title of 'empiricists,' have been chiefly employed in explaining away the obvious fact of experience." The little-known book by Griffin (1998) gracefully eased me into these deep philosophical waters.

8. Throughout this book, I use "intrinsic" to refer to what is internal to the subject, not in the technical sense that analytical philosophers use this term (see, e.g., Lewis, 1983). Unlike mass and charge, which are intrinsic properties of elementary particles, consciousness depends for its existence on background or boundary conditions, for instance, on a beating heart.

9. Nagel (1974). Furthermore, it might even be possible for an intelligent being, such as a computer, to abduce the existence and properties of experiences based on such a rigorous objective theory of consciousness even if it had no experiences itself.

10. Visual and olfactory components jointly create a single experience. However, I can also shift selective attention from the visual modality to the olfactory one, in which case I have a visually-dominated experience followed by an olfactory-dominated one.

11. The physiology of this urge, called micturition, is explored by Denton (2006). His book traces the historical origin of consciousness to instinctual behaviors relating to the absolute essential need for air—nothing invokes a more urgent and powerful somatic reaction than an inability to breathe—as well as to other life-supporting functions, such as pain, thirst and hunger, salt, and micturition.

12. *Zen and the Art of Consciousness* by the psychologist Susan Blackmore (2011) is a pithy and engaging travelogue to the center of her mind by way of meditation. She ends up doubting her very existence, concluding "There is nothing it is like to be me," "I am not a persisting conscious entity," "Seeing entails no vivid mental pictures or movie in the brain," and "There are no contents of consciousness." While I have respect for her valiant attempts to plumb phenomenology, they vividly demonstrate the limit of introspection and why so much philosophy of mind is barren. Evolution has not equipped the mind to access most brain states. We can't introspect our way to a science of consciousness.

13. Forman (1990a), p. 108. The monk Adso uses Eckhartian language to powerful effect in the coda of Umberto Eco's *The Name of the Rose*: "I shall soon enter this broad desert, perfectly level and boundless, where the truly pious heart succumbs in bliss. I shall sink into the divine shadow, in a dumb silence and an ineffable union, and in this sinking all equality and all inequality shall be lost, and in that abyss my spirit will lose itself, and will not know the equal or the unequal, or anything else, and all differences will be forgotten."

14. By the sixteenth-century monastic yogi Dakpo Tashi Namgyal (2004).

15. Phenomenal integration comes in various forms, such as spatial, temporal, spatio-temporal, low-level, and semantic (Mudrik, Faivre, & Koch, 2014).

16. Nietzsche, in *The Genealogy of Morals*, writes: "There is only a perspective seeing, only a perspective knowing." The philosopher Thomas Metzinger (2003) distinguishes three related concepts: mineness, selfhood, and perspectivalness.

17. The subjective singularity of the present moment is at odds with the unchanging, eternal four-dimensional spacetime-block view of general relativity, in which time is simply another dimension and the past and future are as real as the present. Penrose (2004) discusses the conflict between presentism and eternalism.

18. It is not clear whether experience evolves continuously, as the metaphor implies. In chapter 12 of my 2004 textbook (see also VanRullen, Reddy, & Koch, 2010; VanRullen, 2016) I discuss psychological evidence for the static view, on which each subjectively experienced moment of now is one in a series of discrete *snapshots*, like pearls on a string, with the perception of change superimposed. How long each moment lasts in an objective sense is variable, and is tied to the underlying neurophysiology, for instance, to the duration of a dominant oscillatory pattern. This explains moments of protracted duration reported upon in the context of accidents, falls or other life-threatening events; "when I fell, I saw my life flash before me," or "it took him ages to lift the gun and aim at me" (Noyes & Kletti, 1976; Flaherty, 1999).

19. This is paraphrased from Tononi (2012). A variant is "Every conscious experience exists for itself, is structured, is one out of many, is one, and is definite."

Chapter 2

1. The other universes within the multiverse are not causally accessible to us, but exist beyond our cosmological horizon. They are an extrapolation of known physics. The experiences of others, while nonobservable to me, are at least causally accessible as I can interact with their minds in a multitude of ways to induce different experiences. Dawid (2013) discusses the epistemic status of nonobservable objects within the context of string theory. Another example of nonobservable entities is that of the postulated firewalls in the interior of black holes (Almheiri et al., 2013).

2. Some societies seem to operate by quite different reasoning principles. *Don't Sleep, There Are Snakes* (2008) recounts Daniel Everett's adventures among the Pirahã, a small tribe of indigenous people living in the Amazonian rainforest. Everett, a linguist and former missionary, explains many features of the exceedingly simple culture and language of the Pirahã—which famously lacks linguistic recursion—by what he calls the *immediacy of experience* principle. Only what Pirahã have directly seen, heard, or otherwise experienced, or what a third party who has directly witnessed an event reports, is taken to be real. As nobody alive has seen Jesus, they simply ignore any stories about him, explaining why attempts to convert them to Christianity have been ineffective. On the other hand, what happens in their dreams is accepted as real, rather than imagined, events. The Pirahã's extreme empiricism is compatible with the absence of any creation myth, or fiction, or concepts like great-grandparents (owing to their low life-expectancy, few Pirahã have direct knowledge of such creatures).

3. Seth (2015) argues that both the brain as well as the scientific method itself use abductive reasoning to infer facts and laws about the external world from incomplete and noisy sensory and instrumental data. Likewise, Hohwy (2013) views the brain as employing abductive reasoning to infer facts about the external world and its likely behaviors.

4. Psychophysics, or more general, experimental psychology, originated in the first half of the 19th century with Gustav Fechner and Wilhelm Wundt in Germany. Almost two centuries later on, consciousness remains banned from many neuroscience textbooks and classes that, remarkably, fail to mention what it feels like to be the owner of a brain. Two textbooks that take visual phenomenology serious are Palmer (1999) and Koch (2004). See also Bachmann et al. (2007).

5. The community is broadly divided between early and late onset camps for the timing of a visual experience (chapter 15 in my 2004 textbook; Railo et al., 2011; Pitts et al., 2014, 2018; Dehaene, 2015).

6. Your brain is never in the same state from one trial to the next, even if you are looking at the same image. The synapses and organelles inside the nerve cells of your visual center are in a perennial state of agitation, and their exact values will constantly fluctuate from one millisecond to the next. Thus, the absence of a measurable

all-or-none threshold for perception does not imply that no such a threshold exists. See Sergent and Dehaene (2004); Dehaene (2014).

7. David Marr's monograph *Vision* (1982).

8. There are other confidence measures in usage. In one variant, subjects wager a small amount of money, depending on how confident they are of their response. When these protocols are used appropriately, objective and subjective measures of consciousness complement each other well (Sandberg et al., 2010; Dehaene, 2014).

9. Ramsoy and Overgaard (2004) and Hassin, Uleman, and Bargh (2005) discuss unconscious priming and other forms of subliminal perception. The statistical validity of many of these effects are questionable. Indeed, some of the original findings cannot be replicated or show substantial weaker effect size upon replication (Doyen et al., 2012; Ioannidis, 2017; Schimmack, Heene, & Kesavan, 2017; Biderman & Mudrik, 2017; Harris et al., 2013; Shanks et al., 2013). They are victims of small and statistically underpowered experiments filtered through a fiercely competitive publication process biased to report positive findings. Psychology as a field has woken up to the replication crisis and is addressing it. Particularly promising is the rise of preregistered experiments, in which the exact methods, analysis procedures, number of subjects and trials, rejection criteria, and so on are specified in advance, before collecting the data. Many countries require such a registry for clinical trials of new drugs or other therapeutic interventions that directly affect the lives of patients.

10. Pekala and Kumar (1986).

11. The distinction between phenomenal and access consciousness was introduced by the philosopher Ned Block (1995, 2007, 2011). O'Regan et al. (1999), Kouider et al. (2010), Cohen and Dennett (2011), and Cohen et al. (2016) argue that the informational content of consciousness is small. The limited capacity of our conscious perception, described as 7+/–2 chunks of information or, using different measures, as 40 bits per second, is discussed by (Miller, 1956; Norretranders, 1991). Tononi et al. (2016) and Haun et al. (2017) propose ways of capturing the rich content of experience using innovative techniques that avoid the bottleneck of limited short-term memory.

12. See *The New Unconscious* (Hassin, Uleman, & Bargh, 2005) for some of the best empirical studies of the nonconscious.

13. Koch and Crick (2001).

14. Ward and Wegner (2013) review the blank mind literature. Killingsworth and Gilbert (2010) discuss the curious relationship between happiness and mind blanking.

15. From Woolf's "Moments of Being," from *A Sketch of the Past.*

16. Landsness et al. (2011).

17. Medically, Terri Schiavo's case was uncontroversial. She had brief episodes of automatisms: head turning, eye movements, and the like, but no reproducible or

consistent, purposeful behavior. She showed no brain waves on EEG scans, indicating that her cerebral cortex had shut down. This was confirmed at autopsy (Cranford, 2005).

18. Koch and Crick (2001) and (Koch (2004), chaps. 12–13.

Chapter 3

1. The symposium took place in January 2013 at Drepung Monastery in Southern India. The ebb and flow of the debate between the scholars of two very distinct traditions is lovingly reproduced in the evocatively titled *The Monastery and the Microscope* (Hasenkamp, 2017).

2. Descartes writes in a letter to Plempius in 1638 (page 81 of vol. 3 of his *Philosophical Writings*, ed. and trans. by Cottingham et al. [Cambridge University Press]): "For this is disproved by an utterly decisive experiment, which I was interested to observe several times before, and which I performed today in the course of writing this letter. First, I opened the chest of a live rabbit and removed the ribs to expose the heart and the trunk of the aorta. I then tied the aorta with a thread a good distance from the heart."

3. I give precedence to mammals because of the remarkable structural and physiological similarities between human brains and those of other mammals. This makes inferring consciousness in them easier than in animals with quite different nervous systems, such as a fly or an octopus.

4. Judged by single nucleotides polymorphisms (or SNP) in DNA, the difference between people and chimpanzees is 1.23 percent, compared to around 0.1 percent difference in SNPs between two randomly picked humans. However, there are also about 90 million insertions and deletions between the genomes of the two species, for a total of 4 percent variation. These differences built up over the five to seven million years since the last shared ancestor (Varki and Altheide, 2005). For the descent of mammals from a furry creature that lived about 65 million years ago, see O'Leary et al. (2013).

5. Cortex encompasses 14 million neurons in the mouse and 16 billion in humans (Herculano, Mota, & Lent, 2006; Azevedo et al., 2009). Despite this thousand-fold difference, cortical neurons account for one out of every fifth neurons in the brain of both species.

6. I put this to the test at an "All Staff" meeting at the Allen Institute for Brain Science, by asking the couple of hundred employees in attendance to vote (via a phone applet) on which of twelve cortical neurons were human and which mouse (fig. 3.1). I removed the scale bar as human neocortex is 2–3 mm thick while the mouse neocortical sheet is under 1 mm thin and so their total length would be an obvious clue. The audience did barely better than random guessing. The point is not that people can't be trained to distinguish cells from these two species (I'm sure they can) but that

neuronal morphologies are remarkably conserved in both species despite the fact that their last common ancestor lived about 65 million years ago (O'Leary et al., 2013).

7. To compare brains among different species, neuroanatomists invented the encephalization quotient as the ratio of the mass of the brain relative to a standard brain belonging to the same taxonomic group. On this scale, the human brain is 7.5 times bigger than the brain of a typical mammal weighing as much as we do, with all other mammals having smaller encephalization quotients. Why the size of say, the prefrontal cortex, should relate to the size of the body, is left unexplained. With this scale in place, humanity makes it to the top! However, it was recently discovered that long-finned pilot whales, a type of dolphin, have more than 37 billion cortical neurons, compared to the 16 billion cortical neurons in humans. The implications of this for intelligence let alone consciousness remain unknown (Mortensen et al., 2014). I shall return to this theme at the end of chapter 11.

8. Grieving animals have been studied in considerable detail (see, e.g., King, 2013).

9. There are all sorts of interesting differences in the way humans and other mammals see the world. One of the best explored one is color. Almost all mammals—including color-blind men—perceive colors using two types of wave-length sensitive cone photoreceptors (*dichromacy*). Apes and people access a richer color pallet based on three sets of cones (*trichromacy*). Given the genetics of color vision, some women even carry genes for four distinct photo-pigments (*tetrachromacy*). They see subtle colors distinctions that the rest of us are blind to; it remains unclear whether their brain takes advantage of the additional spectral information (Jordan et al., 2010).

10. Whether Neanderthals, Denisovans, and other extinct hominins had conscious experiences is usually not discussed in this context.

11. Macphail (1998, 2000) cites childhood amnesia, the inability of adults to explicit recall events from their childhood prior to 2–4 years of age, as proof of his radical conjecture. According to him, as both animals and babies lack a sense of self and language, they do not experience anything. An even more radical view on consciousness, profoundly ignorant of biology and evolution, that presupposes language, was taken by Julian Jaynes, a brilliant, unconventional sometimes-scientist, whose preposterous ideas continue to enjoy great popularity (for instance, in the popular sci-fi series *Westworld*). In *The Origin of Consciousness in the Breakdown of the Bicameral Mind*, Jaynes (1976) argues that consciousness is a learned process that originated in the second millennium BCE when people discovered that the voices in their heads were not the gods speaking to them but their own internalized speech. The reader is supposed to believe that until that point in time, everybody on the planet was a zombie. Jaynes's prose is elegant, rich in metaphors, full of interesting archaeological, literary, and psychological asides, yet devoid of any brain science or testable hypotheses. Its central thesis is utter nonsense. Greenwood (2015) gives an even-keeled intellectual history.

12. Nichelli (2016) summarizes the extensive academic literature on consciousness and language.

13. Bolte Taylor (2008). For an interpretation of her experiences, in particular with respect to self-consciousness, see Morin (2009) and Mitchell (2009). Marks (2017) is an equally eloquent first-person description of life without an inner voice.

14. Lazar et al. (2000), p. 1223.

15. For reviews of the commodious literation, see chapter 17 in Koch (2004) as well as Bogen (1993) and Vola and Gazzaniga (2017). For a dissenting view on split-brain patients, see note 3 in chapter 10. In almost all right-hand subjects, the dominant, speaking hemisphere is the left one. For lefties, things are a bit more complicated; in some it remains the left hemisphere, in others it is the right, while some have no strong laterality. For the sake of simplicity, I assume throughout this book that the linguistic dominant "speaking" hemisphere is the left one. Bogen and Gordon (1970) and Gordon and Bogen (1974) discuss the involvement of the right hemisphere in singing.

16. The best documented feral child is Genie, a girl in modern-day Los Angeles who was physically restrained, malnourished, and kept isolated by her father until she was discovered as a teenager. Reading her case leaves one numb about such unspeakable evil acts carried out on helpless children (Curtiss, 1977; Rymer, 1994; Newton, 2002).

17. This striking phrase is from Rowlands (2009).

Chapter 4

1. Johnson-Laird (1983) and Minsky (1986). Bengio (2017) updates this metaphor to deep convolutional networks.

2. This attitude is best epitomized by Morgan's canon from his fin-de-siècle textbook *An Introduction to Comparative Psychology* (1894): "In no case may we interpret an action as the outcome of the exercise of a higher psychical faculty, if it can be interpreted as the outcome of the exercise of one which stands lower in the psychological scale." For background, I recommend the *Stanford Encyclopedia of Philosophy*'s entry entitled "Animal Consciousness."

3. See Jackendoff's book *Consciousness and the Computational Mind* (1987), as well as Jackendoff (1996) and Prinz (2003).

4. "In psychoanalysis there is no choice but for us to assert that mental processes are in themselves unconscious, and to liken the perception of them by means of consciousness to the perception of the external world by means of sense-organs" (Freud, 1915, p. 171); or "It dawns upon us like a new discovery that only something which has once been a perception can become conscious, and that anything arising from within (apart from feelings) that seeks to become conscious must try to transform itself into external perception" (Freud, 1923, p. 19). Freud was among the first to

explore the vast dominion of the unconscious, the subterranean source for much of our emotional life. If you have lived through a troubled romantic relationship, you are well acquainted with the maelstrom of love and hope, grief and passion, resentment and anger, fear and desperation, that threaten to suck you in. Spelunking the caverns of your own desires and motivations, rendering them explicit, and thereby possibly understanding them is difficult, for they are consigned to the dark cellars of the mind, where consciousness does not cast its prying light.

5. Crick and Koch (2000); Koch (2004), chap. 18.

6. Hadamard (1945); Schooler, Ohlsson, and Brooks (1993); Schooler and Melcher (1995).

7. See Simons and Chabris (1999). Simons and Levin (1997, 1998) studied many real-life examples of such attentional lapses. Moviegoers usually fail to notice all but the most obvious continuity errors (Dmytryk, 1984). Inattentional blindness and change blindness are two further striking instances of not seeing events or objects that are in full view, demonstrating the limits of perception (Rensink et al., 1997; Mack & Rock, 1998; O'Regan et al., 1999).

8. A popular technique to suppress images from conscious view is continuous flash suppression (Tsuchiya & Koch, 2005; Han, Alais, & Blake, 2018). Using this technique, Jiang and others (2006) had volunteers look at invisible pictures of nude men and women. Attention without consciousness has been demonstrated in numerous experiments that manipulate bottom-up as well as top-down spatial, temporal, feature-based, and object-based attention (Giattino, Alam, & Woldorff, 2017; Haynes & Rees, 2005; Hsie, Colas & Kanwisher, 2011; Wyart & Tallon-Baudry, 2008).

9. Bruner and Potter (1964); Mack and Rock (1998); FeiFei et al. (2007). Dehaene et al. (2006) and Pitts et al. (2018) argue that experience requires selective attention.

10. Braun and Julesz (1998); Li et al. (2002); FeiFei et al. (2005); Sasai et al. (2016). I shall return to the common example of driving while listening in chapter 10.

Chapter 5

1. "Footprints left in the brain" is a particularly apt metaphor for what my colleagues and I are searching for, as it implies that the agent responsible for the footprints remains invisible and has to be abduced.

2. The opening chapter of Gross (1998) overviews classical antiquity's contribution to brain science.

3. Aristotle believed that the brain was essential to the proper functioning of the body and the heart but was subvenient to the heart (Clarke, 1963). The Aristotle quote is from *Parts of Animals*, 656a. He continues, "The brain cannot be the cause of any of the sensations, seeing that it is itself as utterly without feeling as any one

of the excretions.... It has, however, already been clearly set forth in the treatise on Sensation, that it is the region of the heart that constitutes the sensory centre."

4. Zimmer's (2004) book-length account of Thomas Willis and the civil-war-torn England of the seventeenth century rescues the reputation of the founder of neurology from obscurity. The anatomical drawings were made by the young, not yet famous architect Christopher Wren.

5. Science as a distinct activity, with its own professional ethos, methods, and lines of demarcation from pseudo-science, technology, philosophy, and religion, can also be dated to this period, as well as the term "scientist" (first articulated by William Whewell in 1834). See the thick account by Harrison (2015).

6. The word "neuron" wasn't coined until 1891, by the German histologist Wilhelm von Waldeyer-Hartz, to denote the cellular units of the brain. Although neurons get all the press, about half of the brain's cells are not neurons but other cells, including glial cells (astrocytes and oligodendrocytes), immune cells (microglia and perivascular macrophages), and blood-vessel-associated cells (smooth muscle cells, pericytes, and endothelial cells). Collectively, they are less diverse than their neuronal counterparts (Tasic et al., 2018).

7. I here refer only to the chemical synapses that dominate the central nervous system. Electrical synapses provide a direct low-resistance pathway between neurons. They are less common in the adult cortex. For details, see my biophysics textbook (Koch, 1999).

8. Imagine what extraordinary meaning we would attach to these nightly mental excursions if we only dreamt once or twice in a lifetime!

9. The transition from wake to light sleep can occur quite abruptly. It manifests itself in slow drifting movements of the eyes, which are no longer distinguishable from considerably slowed-down saccades, and in a change of the EEG from high-frequency, low-voltage waves to a slower pattern of higher voltage sharp waves in non-REM sleep. In monkeys, one of the manifestation of such pendular deviations of the eyes, changes of the EEG pattern and presumed loss of consciousness is a dramatic and abrupt cessation of firing of a population of "omnipause neurons" in the brainstem with a transition time under one millisecond (Hepp, 2018). These experiments need to be followed up as they might reveal a sharp phase transition that the system is undergoing at this point in time.

10. The device and its successful field test are described in Debellemaniere et al. (2018). See Bellesi et al. (2014) for the underlying science. A challenge for all such noninvasive brain-machine interfaces are complications that arise from electrical currents created by eye, jaw, and head movements, sweating, and loose electrodes. Machine learning is proving to be useful here.

11. Up to 70 percent of non-REM awakenings yield reports of dream experiences (Stickgold, Malia, Fosse, Propper, & Hobson, 2001). Conversely, in a consistent minority of cases, subjects deny having had any experience when awakened from REM sleep. Thus, unlike wakefulness, sleep can be associated with either the presence or absence of conscious experiences. In addition, experiences in dreams can assume many forms, ranging from pure sights and sounds to pure thought, from simple images to temporally unfolding narratives. In summary, it is not easy to determine dreaming by assessing traditional EEG features (Nir & Tononi, 2010; Siclari et al., 2017).

12. Schartner et al. (2017) indeed found that magnetoencephalography (MEG) signals were more diverse during ketamine-, LSD-, and psilocybin-induced hallucinations than during waking, using the Lempel-Ziv complexity score (see chapter 9).

13. Crick's death affected me greatly in expected and unexpected ways. I spoke about this in "God, Death and Francis Crick" for a popular Moth radio hour and in a book chapter (Koch, 2017c). Crick's short monogram, *The Astonishing Hypothesis* (1994), remains one of the best introduction to the big questions of brain science. Ridley (2006) captures Crick's personality well. See also Crick's lively autobiography, *What Mad Pursuit* (1988).

14. Crick and I formulated an empirical program to search for the neural correlates of visual awareness (Crick & Koch, 1990, 1995; Crick, 1994; Koch, 2004). The philosopher David Chalmers was the first to define the NCC in a more rigorous manner (Chalmers, 2000). Over the years, the deceptively simple concept of "neural correlates of consciousness" has been dissected, refined, extended, transmogrified, and dismissed. For all the gory details, I recommend the excellent edited volume by Miller (2015). In this book, I use the operational definition from Koch et al. (2016). Owen (2018) interprets the NCC within a framework informed by Thomas Aquinas's human ontology and Aristotelian metaphysics of causation.

15. As a neuroscientist working with the brain every day, it is easy to slip into an attitude that views neurons as the primary agents causing consciousness, with its dualistic overtones (Polak & Marvan, 2018).

16. Of the 30 trillion cells in a 70 kg adult body, 25 trillion are erythrocytes. Fewer than 200 billion cells, under 1 percent, make up the brain, half of which are neurons. The same body also plays host to about 38 trillion bacteria, its *microbiome* (Sender, Fuchs, & Milo, 2016).

17. Kanwisher, McDermott, & Chun (1997); Kanwisher (2017); Gauthier (2017).

18. Rangarajan et al. (2014), p. 12831.

19. When the left fusiform area was stimulated, facial distortions were absent or were restricted to simple, non-face perceptions, such as winkling and sparkling, traveling blue and white balls, and flashes of light (Parvizi et al., 2012; Rangarajan et al., 2014; see also Schalk et al., 2017). In two more recent studies with epilepsy patients

(Rangarajan & Parvizi, 2016), the left-right asymmetric response to electrical stimulation was inverted. These studies emphasize the import of the mantra *correlation is not causation*. Just because a region becomes active in response to a sight, a sound, or an action (correlation) does not imply that this region is necessary for that sight, sound, or action (causation).

20. Some patients become face-blind following a stroke in the neighborhood of the fusiform face area in cortex (Farah, 1990; Zeki, 1993). Others are face-blind from childhood onward, unable to pick out their spouse at the airport. They are uncomfortable around groups of people, which can come across as shyness or aloofness but is neither. Oliver Sacks, who was face-blind, always asked me to meet him at his apartment, rather than at a restaurant filled with people, as he wanted to avoid embarrassing situations (Sacks, 2010).

Chapter 6

1. Bahney and von Bartheld (2018).

2. Vilensky (2011); Koch (2016b).

3. These nuclei (fig. 6.1) project either directly or via one-hop intermediaries in the basal forebrain to the hypothalamus, the reticular nucleus, and the intralaminar nuclei of the thalamus, and to cortex (Parvizi & Damasio, 2001; Scammell, Arrigoni & Lipton, 2016; Saper & Fuller, 2017). Think of these brainstem nuclei as switches. In one setting, the brain is awake and capable of sustaining consciousness; in a second setting, the body is asleep while parts of the brain remain active. In a third setting, neurons in cortex wax and wane periodically between active and inactive states—a hallmark of deep sleep.

4. Of the 86 billion neurons in the brains of four elder Brazilian men, 69 billion are in the cerebellum and 16 billion in the cortex (von Bartheld, Bahney, & Herculano-Houzel, 2016; see also Walloe, Pakkenberg, & Fabricius, 2014). All remaining structures—thalamus, basal ganglia, midbrain, and the brainstem—account for about 1 percent of all neurons. Female brains have, for unknown reasons, on average, 10–15 percent fewer neurons than male ones.

5. Cerebellar damage can cause nonmotor deficits, resulting in what's known as the cerebellar cognitive affective syndrome.

6. Yu et al. (2014) imaged the brain of the woman born without a cerebellum. A journalistic account of what it is like to grow up without a cerebellum follows the eleven-year-old American boy Ethan Deviney and was published in the 2018 Christmas edition of the *Economist* as "Team Ethan—The family of a boy without a cerebellum found out how to take its place." For other cases of cerebellar agenesis, see Boyd (2010) and Lemon and Edgley (2010). Dean et al. (2010) model cerebellar function as an adaptive filter.

7. It is only recently, with full-brain reconstructions in mice, that the vast purview of pyramidal cell axons is coming into view, with each individual axon connecting, via side branches, to many far-flung regions. In a mouse brain that fits comfortably into a sugar cube, 10 millimeters on each side, the total wiring length of individual axons can exceed 100 millimeters, a widespread web of whisper-thin threads (Economo et al., 2016; Wang et al., 2017a,b). Extrapolating this to the human brain, which is a thousand times more commodious, implies that individual cortical axons are up to one meter in length, with thousands of side branches, each innervating dozens of neurons.

8. Farah (1990) and Zeki (1993) thoroughly review cortical agnosia. The two patients with loss of color perception are described in Gallant et al. (2000) and von Arx et al. (2010). Tononi, Boly, Gosseries, and Laureys (2016) discuss consciousness and anosognosia. The patient who did not see motion, *cerebral akinetopsia* (but whose perception of motion in her auditory and tactile modalities was unaffected), is described in Heywood and Zihl (1999). For a literary feast of the highest order, I recommend the books of the late neurologist Oliver Sacks on this topic.

9. In anosognosia, the brain region that mediates a particular class of experiences, here of spectral colors, is destroyed, and with it, the concrete knowledge of what colors are (except in an abstract sense—you know that bats have sonar to detect prey without knowing what that feels like).

10. The "hot zone" diction was introduced by Koch et al. (2016). Much of the older causal clinical evidence is conveniently forgotten in favor of more recent imaging data that is mere correlative. For this clash, see Boly et al. (2017) and Odegaard, Knight, and Lau (2017). Note that the King et al. (2013) experiments from the group of Stan Dehaene emphasizes the importance of posterior cortex in severely brain injured patients. An ongoing adversarial collaboration involving many protagonists in this debate seeks to resolve these two disparate views using an agreed-upon pre-registered protocol (Ball, 2019).

11. How much to remove is a major dilemma for the surgeon. Cut too much and the patient may become mute, blind, or paralyzed. Cut too little and tumorous tissue may remain or seizures will continue (Mitchell et al., 2013).

12. Prefrontal cortex is here defined as frontal granular cortex—Brodmann areas 8–14 and 44–47—and agranular anterior cingulate cortex (Carlen, 2017). The large size of prefrontal cortex singles out *Homo sapiens* among all primates (but see Passingham, 2002).

13. As a rule, higher mental faculties, such as thinking, intelligence, reasoning, moral judgment, and so on, are more resistant to brain damage than lower-level, physiological functions such as speech, sleep, breathing, eye movements, or reflexes, all of which are associated with specific circuits. This rule, based on a century of neurosurgical practice, is often ignored by brain imagers (Odegaard, Knight, & Lau,

2017). Henri-Bhargava et al. (2018) summarize the clinical assessment of prefrontal lesions. The Sacks reference is to Sacks (2017).

14. The case history of the iconic frontal lobe patient Joe A. is laid out in meticulous detail in a book by Brickner (1936). When A. died nineteen years later, his autopsy confirmed that the surgeon had indeed removed Brodmann areas 8–12, 16, 24, 32, 33, and 45–47, sparing only area 6 and Broca's area so that the patient could speak (Brickner, 1952). The quote is from page 304.

15. Patient K.M. had a near-complete bilateral prefrontal resection for epilepsy surgery (including bilateral Brodmann areas 9–12, 32, and 45–47), after which his IQ improved (Hebb & Penfield, 1940).

16. To better appreciate the state of neurology before brains could be routinely imaged and white and gray matter structures distinguished in live patients (something not possible with x-rays technology of the day), I recommend the autobiographical *A Journey Round My Skull* by the Hungarian author, playwright, and journalist Frigyes Karinthy (1939). He viscerally describes events leading up to and including neurosurgery to remove a large tumor from his brain—while he was fully conscious.

17. Mataro et al. (2001) tell this subject's story. Another example is a young woman with massive bilateral prefrontal damage. While manifesting deficient scores in frontal lobe tests, she did not lose perceptual abilities (Markowitsch & Kessler, 2000). For a striking but less well-documented case, search YouTube for "the man with half a head." Carlos Rodriguez had a car accident around age twenty-seven, with a pole hitting his head, requiring surgical craniotomy. As he did not follow up with skull repair, the top of his skull remains missing. He expresses himself fluently though not always coherently. There is no doubt that he is conscious.

18. Yet the clinical data on the necessity of anterior prefrontal cortex (Brodmann area 10) for metacognition is ambiguous (Fleming et al., 2014; Lemaitre et al., 2018).

19. Hemodynamic activity, measured by a magnetic scanner, in frontal and parietal cortices, correlates with conscious visual (Dehaene et al., 2001; Carmel, Lavie, & Rees, 2006; Cignetti et al., 2014) and tactile perception (Bornhövd et al., 2002; de Lafuente and Romo, 2006; Schubert et al., 2008; Bastuji et al., 2016). Global neuronal workspace theory postulates a nonlinear ignition of frontoparietal networks that enables the emergence of perceptual awareness (Dehaene et al., 2006; Del Cul et al., 2009). However, more refined experiments suggest that these regions are involved in pre- or postexperiential processes, such as attentional control, tasks setting, computing confidence of a judgment, and so on (Koch et al., 2016; Boly et al., 2017), rather than in experience *per se*.

20. This was the theme of my first book on consciousness (Koch, 2004). Tononi, Boly, Gosseries, and Laureys (2016) discuss the more recent literature that supports the hypothesis that primary visual, auditory, and somatosensory cortices are not content-specific NCC.

21. High-resolution brain-imaging techniques are picking up structural differences between the front and the back of cortex (Rathi et al., 2014). A recent structural imaging study implicates regions of human inferior parietal and posterior temporal cortex and precuneus as most distinct from any region in the macaque brain (Mars et al., 2018). The last section in chapter 13 discusses the map-like cortical regions in posterior cortex.

22. Selimbeyoglu and Parvizi (2010) overview the clinical literature concerning electrical brain stimulation (EBS). For the underlying science, including the astonishing degree of spatial specificity of EBS, see Desmurget et al. (2013). Winawer and Parvizi (2016) quantitatively link visual phosphenes to focal electrical stimulation of primary visual cortex in four patients while Rauschecker and colleagues (2011) do something similar for induced motion perceptions in human posterior inferior temporal sulcus (area MT+/V5). Beauchamp et al. (2013) induce visual phosphenes in the temporoparietal junction. Note 18 in chapter 5 and Schalk et al. (2017) discuss face perception following EBS close to the FFA, and Desmurget et al. (2009) the feeling of wanting (intention) to move upon stimulation of the inferior parietal cortex. A pioneering instance of cortical EBS to help a blind volunteer is described in Schmidt (1998).

23. The Penfield and Perot (1963) monograph describes sixty-nine case studies of experiential responses, either due to electrical stimulation of the temporal lobe or during the patient's habitual seizure patterns (most commonly temporal lobe seizures). Two typical spontaneous descriptions of patients upon electrical brain stimulation (EBS) are "it was like being in a dance hall, like standing in the doorway—in a gymnasium-like at the Kenwood Highschool" (case 2 on p. 614), and "I heard someone speaking, my mother telling one of my aunts to come up tonight" (case 3 on p. 617). Repeating the stimulation usually elicited the same response from the patient. There are rare EBS reports from sites in the middle (one patient) and inferior (another patient) frontal gyrus that evoke complex visual hallucinations (Blanke et al., 2000).

24. In Fox et al.'s (2018) hands, about one-fifth of electrical brain stimulation trials of the posterior, but not the anterior, portion of orbitofrontal cortex in epileptic patients elicit olfactory, gustatory, and somatosensory experiences. Popa et al. (2016) trigger intrusive thoughts with EBS in the dorsolateral prefrontal cortex and its underlying white matter in three patients.

25. Crick and Koch (1995) and Koch (2014). The widespread bidirectional connectivity between the claustrum and cortex is quantified by Wang et al. (2017a,b). A triptych of recent papers clarified the role of claustrum neurons in suppressing cortical excitation (Atlan et al., 2018; Jackson et al., 2018, Narikiyo et al., 2018). The spectacular anatomy of claustrum neurons is illustrated in Wang et al. (2017b; Reardon, 2017).

26. These numbers are for 1 mm^3 of mouse cortical tissue (Braitenberg & Schüz, 1998).

27. Takahashi, Oertner, Hegemann, and Larkum (2016) relate the presence of calcium-events in the distal dendrites of layer 5 pyramidal neurons in cortex of mice to their ability to detect a small deflection of their whiskers (see also Larkum, 2013).

28. Starting with von Neumann's comment, "Experience only asserts something like: an observer has made a certain (subjective) perception, but never such as: a certain physical quantity has a certain value," in his 1932 textbook of quantum mechanics (see also Wigner, 1967). For an accessible article describing the problem of defining the boundary between the quantum and the classical worlds, see Zurek (2002).

29. Penrose's 1989 book *The Emperor's New Mind* is a pleasure to read. He answered his critics in a follow-up book, *Shadows of the Mind* (1994). Penrose also invented impossible figures and Penrose tiling. The anesthesiologist Stuart Hameroff added some biological flesh to the quantum gravitational bones of Penrose's theory (Hameroff & Penrose, 2014). It remains highly speculative and vague as to details, without much empirical support (Tegmark, 2000; Hepp & Koch, 2010). Simon (2018) proposes a more concrete and testable hypothesis, based on entangled spins and photons.

30. Photosynthesis in marine algae at room temperatures gains its efficiency by quantum-mechanical electronic coherency within proteins (Collini et al., 2010). There is also evidence to suggest that birds navigate along the Earth's magnetic field using long-lived spin coherences in proteins sensitive to blue light (Hiscock et al., 2016). Both of these effects take place in the periphery. There is currently no credible evidence for quantum entanglement within or across neurons in more central structures, say, in cortex. Unless such data emerges, I remain skeptical (Koch & Hepp, 2006, 2010). Gratiy et al. (2017) discuss the electro-quasi-stationary approximation of the Maxwell's equations that captures electrical events within and across neural circuits at a time scale relevant to experience.

Chapter 7

1. Leibniz's mill argument is from his concise yet cryptic *Monadology*, published in 1714. The quote is from a letter Leibniz wrote in 1702 to Pierre Bayle. See Woolhouse and Francks (1997), p. 129.

2. I recommend Chalmers's (1996) engaging book for his conceivability arguments (see also Kripke, 1980). Shear's (1997) edited volume documents the ripples that Chalmers's formulation of the hard problem left within the philosophical community.

3. The hard problem arises when trying to move from mechanism, say particular types of neurons firing in a particular manner, to experience. IIT takes the opposing and less traveled road—starting with the five indisputable properties of any experience to infer something about the mechanism that constitutes this experience.

4. In contrast to the *deductions* made from axioms in mathematics.

5. On the order of 10^{500}, if we are to believe cosmology's theory of eternal inflation.

6. An edited volume that deals with the origin of the ultimate laws of physics is Chiao et al. (2010). An extreme variant of the multiverse explanation is Max Tegmark's *Mathematical Universe* hypothesis that anything that can exist mathematically must exist physically, somewhere, sometime. All these learned speculations, including on the ultimate question, "Why is there anything at all," end up with variants of "It's turtles all the way down" answers.

7. I closely follow the exposition, examples, and figures from the foundational paper describing IIT 3.0 (Oizumi, Albantakis, & Tononi, 2014). Tononi and Koch (2015) provide a gentle introduction to IIT. See http://integratedinformationtheory.org/ for PyPhi, an open-source Python library for calculating integrated information and an up-to-date reference implementation (Mayner et al., 2018). IIT 3.0 is unlikely to be the final word. There are a number of choices made when turning the five transcendental axioms into five postulates, for instance, which metric to use. These choices will ultimately have to be theoretically grounded. Keep an eye on the growing secondary literature, such as the special issue of *Journal of Conciousness,* vol. 63 (2019).

Chapter 8

1. The critical role of causality in analyzing the web of cause and effect is an emergent theme in biology, network analysis and artificial intelligence. The latter owns much to the foundational work of computer scientist Judea Pearl (2000). I warmly recommend his very accessible *The Book of Why* (Pearl 2018).

2. The axiom of intrinsic existence turns Descartes's epistemological claim, "I know that I exist because I am conscious," into an ontological one, "Consciousness exists," with the supplemental intrinsic claim of "Consciousness exists for itself." For ease of exposition, I compress these into one (Grasso, 2019).

3. The quote is from Plato's *Sophist,* written in 360 BC, line 247d3, available at Project Gutenberg. Equating existence with causal power is known as the Eleatic principle. Note that Plato's disjunctive ("or") requirement is replaced by the stronger conjunctive ("and") requirement in IIT. The causal interaction must flow in both directions, to *and* from the system in question.

4. For simplicity of exposition, the trivial circuit of figure 8.2 is deterministic. The mathematical apparatus is extendable to probabilistic systems to account for indeterminism due to thermal or synaptic noise. It is nontrivial to extend IIT to continuous dynamical systems (such as those described by electrodiffusion or electrodynamics, of relevance to the biophysics of the brain) as partitioning and computing entropy in continuous systems leads to infinities. For one promising attempt, see Esteban et al. (2018).

5. Any graph with *n* nodes can be cut in many different ways into two parts, into three parts, into four parts, and so on. The total number of possible partitions, up

to the one that fully shatters the systems into its individual, atomic components, is elephantine, specified by the nth Bell number B_n. Computing integrated information across all of these partitions is exceedingly expensive and scales factorially. While $B_3 = 5$, B_{10} is already 115,975. For $n = 302$, the number of cells in the nervous system of the worm *C. elegans*, the number of partitions is an absurdly large 10^{457} (Edlund et al., 2011). With some clever mathematics, these numbers can be dramatically reduced. Of course, nature doesn't need to explicitly evaluate all of these cuts to find the minimal one, just as nature doesn't explicitly compute all possible paths a light beam takes to find the one that minimizes its action.

6. In particular, it is not measured in bits, as integrated information is very different from *Shannon information*, as I just explained.

7. The exclusion postulate also solves panpsychism's combination problem, a theme I will return to in chapter 14.

8. The best-known example of such an extremum principle, dating to the seventeenth century, is the one by Pierre de Fermat for optics. His principle of least time states that light travels between two given points along the path of shortest time. It describes how light rays are reflected off mirrors and are refracted through different media, such as water.

9. Rigorously defining what constitutes a Whole and how it is different from the union of its parts is the central occupation of the field of *mereology*. The idea goes back to Aristotle's notions of the soul, form, or essence of any living organism (see his *On the Soul*, also translated as *On the Vital Principle*). Consider a tulip, a bee, and a person. Each of these organisms is made up of multiple organs, structural elements, and interconnecting tissue. The whole has properties—reproduction (all three), movement (the latter two), language (only the last)—not possessed by its parts. The modern emphasis on big data provides the illusion of understanding such systems while obscuring the depth of our ignorance. It remains conceptually challenging to define how these system-level properties come about. Ultimately, maxima in extrinsic causal powers can precisely describe and delineate organisms, such as tulips, bees, and people, while maxima in intrinsic causal powers are essential to experience.

10. Two forms or constellations that differ only by a rotation are the same experience. Mathematicians have begun the exploration of the isomorphism between the geometry of phenomenological space—for instance, the three-dimensional, cylindrical hue, saturation and brightness space for color—and the maximally irreducible cause-effect structure (Oizumi, Tsuchiya, & Amari, 2016; Tsuchiya, Taguchi, & Saigo, 2016). How does the geometry of the form that constitutes the sight of an empty screen, with its spatial extent, differ from that of thirst or boredom?

11. Koch (2012a), p. 130.

12. Note that any difference in causal powers will, of necessity, be paralleled by some difference in the associated physical substrate. That is, a changing experience

is associated with a change in its substrate. This need not be true of the microphysical constituents of this substrate. That is, a neuron might fire one spike more or less which may not affect the physical substrate, and therefore will not change experience, given the particular mapping between microphysical variables and the spatio-temporal granularity relevant for the physical substrate of consciousness (Tononi et al., 2016). Furthermore, because of *multiple realizability*, different physical substrate might instantiate the same conscious experience (see the example in Albantakis & Tononi, 2015). In practice, this is extremely unlikely to happen in brains owing to their massive degeneracy.

13. The famous binding problem of psychology (Treisman, 1996).

14. The Python package PyPhi is described in Mayner et al. (2018) and publicly available, together with a tutorial. The underlying algorithm scales exponentially in the number of nodes. This unfortunately limits the size of the networks that can be fully analyzed, making the ongoing search for heuristics that provide fast approximations to finding the cause-effect structure critical.

Chapter 9

1. Merker (2007).

2. Holsti, Grunau, and Shany (2011).

3. The distressing tale of how the brain can lose its sense of self in late-stage dementia is reviewed by Pietrini, Salmon, and Nichelli (2016). I recommend watching the eloquent and dramatized 2014 movie *Still Alice*.

4. Chalmers's (1998) point was that the construction of a consciousness meter had to await a final, accepted theory of consciousness, a stance I disagree with.

5. Winslade (1998) is a compelling book on traumatic brain injury and the medical ecosystem that enables patients' survival.

6. Posner et al. (2007) is the classic textbook on patients with disorders of consciousness. Giacino et al. (2014) provides an update. There is no central registry for vegetative-state patients, with many relegated to hospices and nursing homes or being cared for at home. Estimates of vegetative-state patients in the United States range from 15,000 to 40,000.

7. Asked to imagine playing tennis or to take an imaginary walk through their house while inside a magnetic scanner, 4 out of 23 VS patients showed the same differential brain responses in their hippocampus and their supplementary motor cortex as healthy volunteers. Related experiments seek to exploit this willful modulation of brain activity as a two-way lifeline for communication ("if the answer is yes, imaging playing tennis"; Bardin et al., 2011; Koch, 2017b; Monti et al., 2010; Owen, 2017).

8. A dramatic exception is the transient, spinal-cord mediated *Lazarus* reflex, in which the corpse raises its arms or even part of its upper body (Saposnik et al., 2000).

9. The original committee of Harvard Medical School faculty issued their report, *Defining Death,* in 1968. Forty years later, another presidential committee reexamined these questions in *Controversies in the Determination of Death.* They reaffirmed the ethical propriety of conventional clinical rules for death—either the neurological standard of total brain failure or the cardiopulmonary standard of irreversible cessation of cardiac and respiratory functions. Several members of the 2008 committee, including its chairman, filed briefs challenging the conclusion that a dead brain implies a dead body. Indeed, there is a handful of reports of brain-dead patients—corpses from the point of view of the law of the land—that, with proper life support, retain the appearance of life for months or years including giving birth to viable babies (Schiff & Fins, 2016; Shewmon, 1997; Truog & Miller, 2014). The book *The Undead* (Teresi, 2012) highlights the conflicting needs of a living body for optimal organ preservation and a dead donor, epictomized by the jarring medical concept of a "beating heart cadaver."

10. Bruno et al. (2016) provide an up-to-date account of LIS. Remarkably, the vast majority of LIS patients want life-sustaining treatment; few confess to suicidal thoughts. Jean-Dominique Bauby's (1997) book, *The Diving Bell and the Butterfly,* is a strangely uplifting and inspirational volume written under appalling circumstances. It was subsequently made into a poignant movie.

11. Niels Birbaumer has dedicated his career to communicating with the worst-off patients, often fully paralyzed, via event-related brain potentials and other brain-machine interfaces. See Kotchoubey et al. (2003) for the science and technology and Parker (2003) for a journalistic account in the *New Yorker.*

12. A good description of how the EEG changes as anesthesia is administered comes from Martin, Faulconer, and Bickford (1959): "Early in the variations from normal comes an increase in frequency to 20 to 30 cycles per second. As consciousness is lost, this pattern of small rapid waves is replaced by a large (50 to 300 microvolts) slow wave (1 to 5 cycles per second) that increases in amplitude as it slows. The wave may become irregular in form and repetition time, and it may have secondary faster waves superimposed as the level of anesthesia deepens. The amplitude next begins to decrease, and periods of relative cortical inactivity (the so-called burst suppression) may appear until the depression finally results in the entire loss of cortical activity and a flat or formless tracing."

13. The original paper is Crick and Koch (1990). For details, see the popular account by Crick (1994) or my textbook (Koch, 2004). Crick and I argued that consciousness required the synchronization of populations of neurons via rhythmic discharges in the gamma range, accounting for the "binding" of multiple stimulus features within a single experience (see Engel & Singer, 2001). Stimulus-specific gamma-range synchronization in cat visual cortex is facilitated by attention (Roelfsema et al., 1997) and by stimulation of the reticular formation (Herculano-Houzel et al., 1999; Munk

et al., 1996). Moreover, gamma synchrony reflects perceptual dominance under binocular rivalry even though firing rates may not change (Fries et al., 1997). Human EEG and MEG studies also suggest that long-distance gamma synchrony may correlate with visual consciousness (Melloni et al., 2007; Rodriguez et al., 1999). The subsequent fate of the Crick–Koch hypothesis is discussed in the following note.

14. Most of these studies confounded selective visual attention and visual consciousness (chapter 4). When the effects of conscious visibility are properly distinguished from those of selective attention, high gamma synchronization is tied to attention, independently of whether or not the stimulus was seen by the subject, while mid-range gamma synchronization relates to visibility (Aru et al., 2012; Wyart & Tallon-Baudry, 2008). Hermes et al. (2015) demonstrated the existence of gamma oscillations in human visual cortex but only when looking at specific types of images. This led to the conclusion that gamma-band oscillations are not necessary for seeing (Ray & Maunsell, 2011). Finally, gamma synchrony can persist or even increase in early NREM sleep, during anesthesia (Imas et al., 2005; Murphy et al., 2011) or seizures (Pockett & Holmes, 2009) and can be present for unconscious emotional stimuli (Luo et al., 2009). That is, gamma synchrony can occur without consciousness.

15. Kertai, Whitlock, and Avidan (2012) describe the pros and cons of using the BIS for anesthesiology.

16. The P3b is a well-studied electrophysiological candidate marker of consciousness. It is a late (>300 ms after the onset of stimulus), positive, frontoparietal event-related potential evoked by visual or auditory stimuli, first described fifty years ago. Measured using an auditory oddball paradigm, the P3b has been proposed as a signature of consciousness, revealing a nonlinear amplification (also referred to as *ignition*) of cortical activity through a distributed network involving frontoparietal areas (Dehaene & Changeux, 2011). However, this interpretation has been contradicted by a variety of experimental findings (Koch et al., 2016). The *visual awareness negativity* (VAN), an event-related potential deflection that starts as early as 100 ms after the onset of a stimulus, peaks around 200–250 ms, and is localized to the posterior cortex, correlates better with conscious perception (Railo, Koivisto, & Revonsuo, 2011).

17. The original study (Massimini et al., 2005) correctly distinguished quiet restfulness from deep sleep in a small group of normal subjects. In the intervening years, Tononi, Massimini, and a large team of clinical associates tested the transcranial magnetic stimulation zap-and-zip procedure in conscious and unconscious states in volunteers and neurological patients. I wrote about this story for a cover article of *Scientific American* (Koch, 2017d). See the up-to-date monograph for further details (Massimini & Tononi, 2018).

18. Faraday or electromagnetic induction, the principle that a changing magnetic field induces voltage in a conductor, is at the heart of electrical generators (or dynamos) and wireless charging.

19. See http://longbets.org/750/. Such technologies must be published in the open, peer-reviewed literature and must be based on hundreds of subjects. The procedures must be validated on individual, clinically unambiguous cases, have a very low miss rate (labeling somebody unconscious when they are conscious) and a low false-alarm rate (labeling somebody conscious when they are not). The *Long Bet* follows an ongoing twenty-five-year bet about the nature of the NCC between David Chalmers and myself (Snaprud, 2018). In the end, purely data-driven methods to distinguish conscious from nonconscious states, using machine learning, might win out (Alonso et al., 2019).

20. For instance, PCI is not monotonic in Φ. Given the normalization for source entropy in the calculus, PCI reaches its maximum for evoked responses in which the sources are completely independent of each other, implying a complete lack of integration across cortex. In practice, PCI does not exceed 0.70.

21. The brain has many organizational levels: voxels of the sort seen in magnetic scanners that include a million or more cells; coalitions of neurons underneath a clinical electrode; individual nerve cells accessible to modern optical or electrical recording techniques; their contact points, synapses; the proteins that are packed into synapses—and on and on. The intuition of most working neuroscientists is that the relevant causal actors are assemblies of discrete nerve cells. Integrated information theory does better than intuition. According to IIT, the neural elements of the NCC are those, and only those, that support a maximum of cause-effect power, as determined by the intrinsic perspective of the system itself. This can be assessed empirically (fig. 2 in Tononi et al., 2016). The same logic dictates the time scale relevant to the NCC as the one that makes the most difference to the system, as determined by its intrinsic perspective. This time scale should be compatible with the dynamics of experience—the rise and fall of percepts, images, sounds, and so on, in the range of a fraction of a second to a few seconds.

Chapter 10

1. There are a number of much smaller fiber bundles besides the corpus callosum, in particular the anterior and posterior commissures, connecting the two cortical hemispheres. A partial or complete *callosal disconnection syndrome* can also be a rare complication of chronic alcoholism (Kohler et al., 2000) or of trauma, as when a Japanese businessman rammed an ice pick into his head while drunk. He walked himself to a hospital with the handle of the pick protruding from his forehead (Abe et al., 1986).

2. Injecting the short-term barbiturate sodium amobarbital into the left carotid artery puts the left hemisphere to sleep. If the patient continues to speak, Broca's area must be in the right hemisphere. Known as the Wada test, this procedure remains the gold standard for determining the lateralization of language and memory function, outperforming fMRI (Bauer et al., 2014).

3. Recent tests of two split-brain patients have questioned the orthodox narrative (Pinto et al. 2017a,b,c; and the critical replies by Volz & Gazzaniga 2017 and Volz et al. 2018). Any interpretation of split-brain patients has to contend with brain reorganization following the surgical intervention that occurred years or even decades earlier. Another puzzling observation are people born without a cortical commissure, so-called agenesis of the corpus callosum, that do not have classical split-brain symptoms. Indeed, activity in their left and right cortices rises and falls together, despite lacking direct structural connections (Paul et al., 2007).

4. Sperry (1974), p. 11. The idea that the left and right halves of the brain are reflected in a duality of mind goes back a long time (Wigan, 1844). The philosopher Puccetti (1973) wrote a fictionalized court case of a patient whose nondominant hemisphere murdered his wife in a particularly lurid manner; see also Stanisław Lem's 1987 satirical science fiction novel *Peace on Earth.*

5. Split-brain patients, after convalescence from their surgical disconnection, claim not to feel significantly different from their pre-operation selves. This needs to be accounted for in terms of their Φ^{max} before and afterward. As these patients suffered from the deleterious effects of long-term epilepsy, their brains are not normal. Furthermore, just because subjects claim not to feel any different across weeks or months of recovery does not mean that there aren't any differences. It would be essential to map the mind of these patients before and after surgery (that is rarely performed today) using the sort of detailed questionnaires mentioned in the second chapter.

6. A closely related phenomenon is alien hand syndrome. A classical account (Feinberg et al., 1992) tells of a patient whose left hand had a will of its own, choking her. It took great strength on her part to pull her left hand off her throat. Another is of a man whose right hand had a prominent grasp response—in constant motion, it groped nearby objects, including bedclothes or the patient's own leg or genitals, and did not release them. Who can forget the iconic scene in Stanley Kubrick's 1964 nightmare comedy *Dr. Strangelove, or How I Learned to Stop Worrying and Love the Bomb* in which Dr. Strangelove's (played by Peter Sellers) black-gloved right hand abruptly gives the Nazi salute and, when his left hand intervenes, tries to throttle him.

7. Check out the *New York Times Magazine* story and the extraordinary video of the two little girls, Tatiana and Krista Hogan, whose brains are connected at the level of their thalami (Dominus, 2011; https://www.cbc.ca/cbcdocspov/episodes/inseparable).

8. A rare real-time experiment in controlling single neurons in the human brain was carried out by Moran Cerf, at the time a graduate student in my laboratory at Caltech, working with the neurosurgeon Itzhak Fried (Cerf et al., 2010). Patients observed the activity of individual neurons in their own medial temporal lobe and willfully up- and down-modulated them (presumably along with the activity of a great many other cells). State of the art recording technology, a single *Neuropixels* silicon probe thinner than a human hair, can record from several hundred neurons at a time (Jun et al.,

2017). We are still a number of years away from being able to simultaneously record, at millisecond resolution, from one million neurons, less than one-hundredth of one percent of cortical neurons, while our ability to selectively stimulate them, to regulate their firing rate up and down, is even more limited.

9. The claustrum with it massive bidirectional connections to and from cortex is an obvious candidate for this role. The *crown of thorns* claustrum neurons spreading their influence far across the cortical mantel might be essential for the establishment of a single dominant Whole (Reardom, 2017; Wang et al., 2017b).

10. Figure 16 in Oizumi, Albantakis, and Tononi (2014).

11. Mooneyham and Schooler (2013).

12. Sasai et al. (2016) describe fMRI evidence for two functionally independent brain networks when subjects have to simultaneously drive a car and listen to instructions, an everyday experience.

13. A wealth of data in the psychiatric, psychoanalytical, and anthropological (e.g., shaman, possession) literature could be mined for insights of direct relevance to (dys)functional brain connectivity (Berlin, 2011; Berlin & Koch, 2009). Another, culturally celebrated, dissociation takes place when falling in love. The attendant reality distortion can occur quite suddenly, and is experienced as highly pleasurable. It frees up enormous physical and creative energies, although it can also lead to maladaptive actions. It too can be viewed through the lenses of IIT.

14. With possibly up to eight separate minds in the case of the octopus (Godfrey-Smith, 2016).

15. I observe this most acutely when stepping onto any transcontinental flight. The first action of the vast majority of passengers is to turn on the in-seat monitor and watch one movie after another, interrupted perhaps by sleep, until the plane arrives, ten or more hours later, at its destination. No desire for introspection or reflection. Most do not want to think, preferring the passive onslaught of images and sounds. Why are so many so ill at ease with their mind?

16. It continues: "(This awareness) is empty and immaculately pure, not being created by anything whatsoever. It is authentic and unadulterated, without any duality of clarity and emptiness. It is not permanent and yet it is not created by anything. However, it is not a mere nothingness or something annihilated because it is lucid and present." From Padmasambhawa's seventh song, in Odier (2005).

17. Each culture interprets these mystical phenomena in a different way, depending on its religious and historical sensitivities. Common to all is an experience devoid of content (Forman, 1990b).

18. What I call *mystical experiences* are quite different from a second class of religious experiences—rapturous, blissful, or *ecstatic experiences*. These are associated with

positive affect and sensory imagery such as the spiritual vision of a Teresa of Avila, the ecstatic experience found in modern Pentecostal or charismatic Christianity, and in the whirling dervishes of Sufism (Forman, 1990c). These two classes represent opposite poles of experience—one has little to no content while the other is overwhelming in its content. Common to both are the endogenous nature of the experience and its long-term influence on the lives of the experiencer that is universally reported to be revelatory and beneficial.

19. Lutz et al. (2009, 2015); Ricard, Lutz, and Davidson (2014).

20. A deflationary explanation that my daughter offered was that I fell asleep. This is, of course, a legitimate concern, which is why I would like to float again while wearing wireless EEG sensors to physiologically distinguish "unconscious while deeply asleep" from "pure experience while awake."

21. There is a paradox here. This state must have some phenomenal aspect to it, as it feels different from deep sleep. Thus, it can only be content free in the asymptotic limit.

22. Sullivan (1995).

23. The mathematics of the inactive cortex is discussed in a toy example in figure 18 of Oizumi, Albantakis, and Tononi (2014). Consciousness depends on appropriate background conditions. Thus, IIT predicts that pure consciousness corresponds to an aroused brainstem (that is, the subject is not asleep and the cortex is suffused with the relevant neuromodulators), and a posterior cortical hot zone that is only minimally active. This prediction can be tested by having long-term meditators wear a high-density EEG cap while engaged in content-less and content-ful meditation, such as focusing on one's breath. Contrasting the EEG signatures of these two conditions should reveal reduced high-frequency, gamma-band activity during content-less meditation, and little delta-band activity.

24. In other words, physicalism is not violated.

Chapter 11

1. Earl (2014) lists the remarkably wide-ranging cognitive functions attributed by scholars to consciousness.

2. Dribbling a soccer ball or typing on a keyboard are well-rehearsed visuomotor skills that suffer if subjects are forced to attend to which side of their foot they are contacting the ball with or which finger they are using to press a particular key (Beilock et al., 2002; Logan & Crump, 2009).

3. Wan et al. (2011, 2012) investigates the development of unconscious skills and their neural correlates for the Japanese strategic board game Shogi. See Koch (2015) for a summary.

4. See note 9 in chapter 2.

5. See the trifecta of papers: Albantakis et al. (2014); Edlund et al. (2011); Joshi, Tononi, and Koch (2013). The mazes are randomized to prevent the evolution of animats specialized to run just one maze. These animats are patterned after the *vehicles* from the eponymous 1986 book *Vehicles: Experiments in Synthetic Psychology* of my PhD co-advisor Valentin Braitenberg. The little disks in figure 11.1 are shaded differently, indicative of when this animat evolved, ranging from light gray for animats that arose early on to black for those digital organisms that developed quite late in evolution.

6. Simulating evolution taught me the immense time scales required for the chance discovery of even simple circuits, such as a one-bit memory, and the degeneracy of the solution space (a very large number of different circuits can accomplish the same function).

7. Albantakis et al. (2014).

8. Crick and Koch (1995), p. 121. See also Koch (2004), chapter 14.1.

9. Differences in general intelligence, assessed in this way, correlate with success in life, social mobility, job performance, health, and lifespan. In a study of one million Swedish men, an increase in IQ by one standard deviation was associated with a 32 percent reduction in mortality over twenty years (Deary, Penke, & Johnson, 2010). My point about the dissociation between cognitive measures of intelligence and consciousness applies to other measures of intelligence, such as those emphasizing social situations, as well. See Plomin (2001) for a mouse intelligence measure.

10. Neuroscience's understanding of the relationship between brain size, complexity of behavior, and intelligence is rudimentary (Koch, 2016a; Roth & Dicke, 2005). How do animals from different taxa, with nervous systems of radically different sizes, deal with conflicting information, say, a red light that can signal food in one context but electric shock in a different one? Can bees learn to deal with such contingencies as well as mice, once differences in motivation and in sensory systems have been accounted for, even though bees have 70 times fewer nerve cells than mice? The cortex of the long-finned pilot whale, a kind of dolphin, contains 37 billion neurons compared to the 16 billion for humans (Mortensen et al., 2014). Are the few thousand individuals that remain of these graceful aquatic mammals really smarter than humans? The cranial capacity of the extinct member of the hominid family, *Homo neanderthal*, was about 10 percent bigger than that of modern *Homo sapiens* (Ruff, Trinkaus, & Holliday, 1997). Were our archaic cousins smarter but less fecund or aggressive than modern humans (Shipman, 2015)? When comparing across species, it is important to tease apart the allometric relationships between body mass, brain mass, and number of neurons in different brain regions.

11. This broad hypothesis is likely to be true only for certain connectivity rules, e.g., for cortex but not for the cerebellum. An interesting mathematical challenge

is to discover the conditions under which integrated information of 2-D networks of simple processing units increases with network size. Will such planar architectures resemble the local, grid-like connectivity, supplemented by sparse nonlocal wiring, that evolution discovered more than 200 million years ago, with the advent of mammals and their neocortical sheet (Kaas, 2009; Rowe, Macrini, & Luo, 2011)?

12. That is, is a person, dog, mouse, or bee with a larger nervous system smarter and more conscious than a person, dog, mouse or bee with a smaller one? While such an idea comes with phrenological and political overtones when studied in the context of humans, the data support such an assertion for intelligence; the thicker the cortical sheet, the higher the IQ score of the subject, as measured by the Wechsler Adult Intelligence Scale (Goriounova et al., 2019; Narr et al., 2006). However, in most species, such as humans, size variations are modest. Owing to selective breeding in dogs, the range in body size across breeds of *canis domesticus*—from Chihuahuas to Alaskan malamutes and Great Danes—is large, at least a factor of one hundred in body mass. It would be interesting to measure the number of neurons in different brain regions across these breeds and to seek to correlate these numbers with performance on some battery of standardized behavioral tests (Horschler et al., 2019).

13. Shinya Yamanaka was awarded the 2012 Nobel Prize for this major discovery in the field of regenerative medicine.

14. The technology of growing brain organoids by reprogramming adult human stem cells is progressing rapidly (Birey et al., 2017; Di Lullo & Kriegstein, 2017; Quadrato et al., 2017; Sloan et al., 2017; Pasca, 2019). Cerebral organoids obviate the need for abortion-derived embryonic tissue that remains ethically controversial and can be grown in very large numbers under tightly controlled conditions. So far, organoids lack microglia and cells that form blood vessels (but see Wimmer et al., 2019). That limits their size to that of a lentil, containing perhaps up to one million cells. To grow any larger, organoids need vascularization to deliver oxygen and nutrients to cells on the inside. The morphological and electrical complexity of these neurons is much less than mature neurons, with limited synaptic connectivity, and irregular neuronal activity that is different from the organized activity patterns characteristic of consciousness that I discussed in figure 5.1 and in chapter 6. A recent landmark study reported prolonged periods of relative quiet, punctuated by episodes of spontaneous electrical activity, including nested oscillations and high variability that somewhat resemble the EEG of preterm infants (Trujillo et al., 2018). Stay tuned.

15. Narahany et al. (2018) discuss the ethics of experimenting with human cerebral organoids. In chapter 13, I argue that extended grids of neurons, such as those that might arise in cerebral organoids, experience something akin to the phenomenology of empty space, with its inherent neighborhood relations and distances. I also argue that cortical carpets are the living repudiation of an interesting objection raised against IIT by the physicist Scott Aaronson.

Chapter 12

1. "I therefore claim to show, not how men think in myths, but how myths operate in men's minds without their being aware of the fact" (Levi-Strauss, 1969), p. 12.

2. Leibniz (1951), p. 51.

3. Among other things, the young Turing proved that Leibniz's desire for a way to answer any properly posed question (see the above Leibniz quote) with a "true" or "false," the notorious *Entscheidungsproblem*, is an unattainable goal.

4. The stream of incoming photons is captured by 6 million cone photoreceptors in the eye, each modulated up to 25 Hz with a signal-to-noise ratio of 100, for a total of about one billion bits per second of information (Pitkow & Meister, 2014). Most of this data is discarded within the retina, such that only about 10 million bits per second of information leave the eye ball, streaming along the one million fibers making up the optic nerve (Koch et al., 2006). For a recent estimate of keyboard and speech processing rates, see Ruan et al. (2017).

5. There is a substantial literature on functionalism. See the recent entry in *The Stanford Encyclopedia of Philosophy* (Levin, 2018). Clark (1989) introduced micro-functionalism.

6. Hubel (1988).

7. Through the Allen Brain Observatory (de Vries, Lecoq et al., 2018).

8. Zeng and Sanes (2017) describe the modern view of the taxonomic classification of brain cells into classes, subclasses, types and subtypes and its similarities and differences to classifying species. Neuronal cell types are best understood in the retina, where there is remarkable little variation in the types of neurons and their circuitry across mammalian species (Sanes & Masland, 2015).

9. Arendt et al. (2016) introduce the developmental and evolutionary constraints underlying cell types.

10. Each organism is the outcome of a long chain of predecessors, all the way back to the origin of life. Features are constantly adapted and reused in novel ways across evolutionary time scales. Take ossicles, the small bones in our middle ear that transmit sound. They evolved from bones in the jaws of early reptiles that can be traced back to gills in even earlier tetrapods. Evolution turned a breathing aid into a feeding aid that morphed into a hearing aid (Romer & Sturges, 1986). See also the lucid *Your Inner Fish* (Shubin, 2008), which explains with panache the evolutionary origin of so many features of our bodies. Similarly, many of the extant cell types are likely to be evolutionary remnants from deep time. Think of the arrangement of the QWERTY keyboard that is such a dominant feature of our life. It is a mechanical constraint imposed on late-nineteenth-century typewriters to minimize the chance of nearby keys physically colliding and jamming. Irrelevant to virtual electronic keyboards, the key layout remains a powerful reminder of the long reach of history.

Some of the cell types of the brain are likely critical for the organism as it self-assembles from a single fertilized egg, by way of a blastocyst, into an embryo that will become a newborn. Development imposes unique design constraints that are not well understood. One example is the retina, which develops from the inside out, explaining why the light-sensing photoreceptors are situated at the back of the eye rather than in the front as in all cameras.

Then there are metabolic constraints. Your brain consumes one-fifth of the power needed by your body at rest—about 20 watts, whether you are watching a movie, playing chess, or sleeping. Eating a medium-sized banana an hour provides enough caloric energy to feed your body and your brain. The metabolic cost of brain tissue is high compared to kidney or liver tissue. Evolution had to come up with ingenious ways to perform low-power operations; it didn't care about computational universality or elegance.

11. Alan Hodgkin and Andrew Huxley's 1952 work remains a high-water mark of computational neuroscience. They inferred the changes in the membrane conductance that underlies the initiation and propagation of the action potential in the squid giant axon using pre-transistor recording equipment. They formulated a phenomenological model that quantitatively reproduced their observations in terms of the interaction between voltage- and time-dependent sodium, potassium, and leakage conductances. It took them three weeks (*sic*) to solve the associated four coupled differential equations on a hand-cranked calculator and derive the numerical value of the propagation speed of the action potential, which was within 10 percent of the observed value (Hodgkin, 1976). I am in awe of their achievements, which netted them the Nobel Prize in 1963. Modified versions of their calculus remains at the core of all realistic neuronal modeling efforts in the field today (Almog & Korngreen, 2016).

12. Henry Markram's Blue Brain Project is funded by the Swiss government (Markram, 2006, 2012). The Blue Brain Project has achieved a first-draft digital reconstruction of a small piece of rat somatosensory cortex (Markram et al., 2015) that is, without doubt, the most complete simulation of a grain of excitable brain matter (for my detailed thoughts, see Koch & Buice, 2015). For more recent simulations, see https://bluebrain.epfl.ch/. If current trends in the hardware and software industry continue, high-performance computing centers will have the raw computational and memory capabilities to simulate the neural dynamics of the rodent brain by the early 2020s (Jordan et al., 2018). A cellular-level human brain simulation, with a thousand times more neurons that are not only more extensive than mice neurons but also carry many more synapses, remains out of reach for now (as of 2019).

13. While, in principle, any computer system can imitate any other, in practice this is very challenging. Emulators are explicitly designed to imitate a *target* computer system on a *host* computer system using either specialized hardware or software (such as microcode). Examples are emulators that run under an Apple Operating System and mimic the look and feel of a Windows environment (or vice versa) or

video game emulators for old console games, such as Super Nintendo, running on contemporary PCs. An important consideration is the execution speed of the emulator compared to the original system.

14. See in particular my books, Koch and Segev (1998) and Koch (1999), and the detailed simulations, in collaboration with the Blue Brain Project, of electric fields in the mammalian brain (Reimann et al., 2013).

15. See Baars (1988, 2002), Dehaene and Changeux (2011), and Dehaene's excellent (2015) book. Empirical support derives from experiments that manipulate stimulus visibility through masking, inattentional, and change blindness, as well as fMRI and evoked electrical potential in humans and neuronal recordings in nonhuman primates (van Vugt et al., 2018).

16. For a criticism of the Global Neuronal Workspace theory from the point of view of IIT, see the appendix in Tononi et al. (2016). While IIT takes no direct position on the relationship between attention and experience, Global Neuronal Workspace theory posits that attention is necessary for access to the workspace. Furthermore, because of the small size of the workspace, the content of consciousness is limited, while there is no such constraint in IIT. An ongoing pioneering experiment in the sociology of cognitive neuroscience, an *adversarial collaboration*, seeks to resolve some key open questions regarding the NCC on which Global Neuronal Workspace and Integrated Information theories differ. The collaboration agreed on a preregistered set of experiments using fMRI, MEG, EEG, and electrodes implanted in epileptic patients (Ball, 2019).

17. See Dehaene, Lau, and Kouider (2017), p. 492, in which they discuss the possibility of machine consciousness within global neuronal workspace theory, concluding that "current machines are still mostly implementing computations that reflect unconscious processing."

Chapter 13

1. Adding "trivial" feedback does not necessarily invalidate this conclusion (Oizumi, Albantakis, & Tononi, 2014). Note that the idealized, abstract feedforward system treated here may well have non-zero Φ^{max} once it is built out of real physical components, given reciprocal interactions at the physical microlevel (Barrett, 2016).

2. A feedforward network can sustain rapid visuomotor action, as when your brain signals within 120 milliseconds that a flashed image contains a threat or when your hand unthinkingly shoots out to catch a glass about to tip over, but without any conscious experience (which may come later, delayed by a fraction of a second). Psychologists and neuroscientists have always emphasized the necessity of feedback for consciousness. See Cauller and Kulics (1991); Crick and Koch (1998); Dehaene and Changeux (2011); Edelman and Tononi (2000); Harth (1993); Koch (2004); Lamme (2003); Lamme and Roelfsema (2000); Super, Spekreijse, and Lamme (2001). Lamme's *recurrent processing* theory of consciousness makes this requirement explicit (Lamme 2006, 2010).

3. The dynamics of the internal processing units of the feedforward network are faster than those of the recurrent one. The key idea in designing this network is to unfold the elements of the recurrent net by passing the state of each element through a chain of four nodes. Functional equivalence can't be proven for input sequences longer than four time-steps (see fig. 21 in Oizumi, Albantakis, & Tononi, 2014).

4. Feedforward networks with a single intermediate (so-called *hidden*) layer containing a finite number of neurons can approximate any measurable function under some mild assumptions (Cybenko, 1989; Hornik, Stinchcombe, & White, 1989). In practice, given the need to learn from millions of curated examples, feedforward nets are deep, i.e., have many hidden layers. See Eslami et al. (2018) for more on visual imagery and imagination in computer vision.

5. Findlay et al. (2019).

6. You can do it yourself or you can consult the appendix of Findlay et al. (2019), where it is done for you. It takes eight updates of the computer to simulate one time-step of the (PQR) network. That is, after eight clock iterations, the computer has simulated the transition of (PQR) from its initial state (100) to its next state (001). After another eight iteration, the simulation correctly predicts that (PQR) will be in state (110).

7. The entire analysis has to be repeated for the remaining seven of the eight steps that the computer needs to simulate one transition of the (PQR) circuit. However, the result is the same. The system never exists as a single Whole, but rather fragments into nine tiny Wholes (Findlay et al., 2019).

8. This circuit implements the famous computational rule 110 (Cook, 2004).

9. This can be proven by induction. Of course, this design principle is clearly not a practical way to simulate a human brain, as it would require on the order of 2^{billions} gates, vastly more gates than atoms in the universe, but the principle still holds.

10. Marshall, Albantakis, and Tononi (2016).

11. Findlay et al. (2019).

12. The philosopher John Searle expressed similar antifunctional and anticomputational sentiments in his Chinese room argument (Searle, 1980, 1997; see also the *Wordstar* discussion in Searle, 1992). He also argued that the causal powers of the brain give rise to consciousness, yet without fleshing out what he means by this. IIT is not only fully compatible with Searle's intuitions but makes his notions precise. In a review of an earlier book of mine, Searle attacks IIT for its supposedly extrinsic usage of information (as in Shannon's information theory; see Searle, 2013a,b, and our reply, Koch & Tononi, 2013). Meetings between Searle, myself, and Tononi have failed to resolve this bizarre misapprehension. Coming from a philosopher who made his reputation on the notion of what it means to understand, this is rich with irony. Fallon (2019b) discusses the affinity between Searle and IIT in considerable detail,

concluding that Searleans should take delight in IIT's central role of causal power and that "IIT fills in potentially damning gaps in Searle's account."

13. Its sparse connectivity matrix would have 10^{11} times 10^{11} entries, most of which are, however, zero.

14. Friedmann et al. (2017).

15. Floating point units (FPUs), graphics processing units (GPUs), and tensor processing units (TPUs) need to be analyzed in the same manner to determine their intrinsic causal power.

16. Turing never intended his imitation game (Turing, 1950) to be used as a test for consciousness, but rather intended it as a test for intelligence.

17. Seung (2012). All contemporary scanning technologies are destructive; that is, your brain would die in the process of obtaining your brain's connectome.

18. See Aaronson's blog *Shtetl-Optimized*, https://www.scottaaronson.com/blog/?p=1823, which also includes Tononi's detailed reply as well those of many others. It makes for fascinating reading.

19. Mathematicians first studied expander graphs in the context of neural networks in which both neurons (vertices) as well as their axons and dendrites (edges) take up physical space and thus can't be packed arbitrarily closely together (Barzdin, 1993).

20. It is no easy matter to compute integrated information for 1-D and 2-D grids of simple cellular automata or logic gates (for one example, see figure 7 in Albantakis & Tononi, 2015). Approximations indicate that for planar grids, $\Phi^{\max}$ scales as $O(x^n)$, where x depends on the details of the elementary logic gate or rule implemented by the cellular automata and n is the number of gates in the grid element.

21. Its Kolmogorov complexity is very low.

22. An elegant demonstration of the topographic nature of human primary visual cortex aligns electrical stimulation of the tissue at specific anatomical locations to the perceived location of induced light flashes, so called phosphenes (Winawer & Parvizi, 2016). For a pointer to the vast brain imaging literature, see Dougherty et al. (2003), or any neuroscience textbook.

Chapter 14

1. In the limit, the ultimate arbiter of consciousness is the experiencing individual.

2. The drawing in figure 14.1 is an artistic rendition of the popular *tree of life* of David Hillis (loosely based on appendix A of Sadava et al., 2011). The prokaryotic lineages are indicated by dashed lines. The four eukaryote groups (brown algae, plants, fungi, and animals) that contain multicellular organisms are designated by solid lines. The

asterisk marks the location of the one species within the mammalian leaf that ferociously believes it is singular. Note that this depiction oversimplifies the relationships among species as it neglects horizontal gene transfer (sideways heredity).

3. One caveat is that not all nervous architectures are equally privileged with respect to consciousness. I discussed in chapter 6 the clinical evidence that partial or complete loss of the cerebellum does not significantly change patients' consciousness. Thus, for species without a cortex, one has to investigate to what extent the wiring of their brain is more like a cortex or more like a cerebellum. For the case of the bee's brain, its anatomical complexity resembles that of the highly recurrent cortex.

4. Barron and Klein (2016) argue for insect consciousness based on similarities between insect and mammalian neuroanatomy. Bees can be trained to fly through mazes using sophisticated cues that they remember (Giurfaa et al., 2001) and deploy selective visual attention (Nityananda, 2016). Loukola et al. (2017) describe social learning in bees. The literature on nonmammalian cognition, communication, and consciousness is vast. Three erudite and compassionate classics I recommend are Dawkins (1998), Griffin (2001), and Seeley (2000). Feinberg and Mallatt (2016) provide a magisterial book-length account of the origin of consciousness following the Cambrian explosion 525 million years ago.

5. Darwin (1881/1985), p. 3.

6. *The Hidden Life of Trees* (Wohlleben, 2016) by a German forester. See also Chamovitz (2012), Gagliano (2017), and Gagliano et al. (2016) for the growing literature on the complex chemical senses and associative learning capabilities of plants.

7. Some biologists, going back to Ernst Haeckel's biopsychism, and philosophers argue that life and mind are coextensive, sharing common principles of organization (Thomson, 2007).

8. As discussed in chapter 8, anything with a global, non-zero maximum of integrated information over the substrate considered (e.g., the brain) is a Whole. That is, there is no superset or subset of the substrate considered with more integrated information. There will, of course, be other nonoverlapping Wholes, such as the brains of others, with more integrated information.

9. Jennings (1906), p. 336.

10. The numbers are from Milo and Philips (2016). The most detailed cell-cycle model of a single-cell organism is that of Karr et al. (2012) of the human pathogen *Mycoplasma genitalium*, with its 525 genes. However, it still leaves out the vast majority of protein-to-protein interactions. Note that unlike the brain, which incorporate learned regularities about the outside world into its synaptic connectivities, proteins diffuse in the intracellular aqueous environment and are unlikely to have the same connectional specificity. I have approximated the full number of dense interactions

in various organisms and made a computational argument that we can never analyze all of them (Koch 2012b).

11. Properly computing the integrated information of elementary particles requires a quantum version of IIT (Tegmark, 2015; Zanardi, Tomka, & Venuti, 2018).

12. I'm perfectly cognizant that the last forty years in theoretical physics have provided ample proof that chasing after elegant theories, such as supersymmetry or string theory, has yielded no new, empirically testable evidence describing the actual universe we live in (Hassenfelder, 2018).

13. Chalmers (1996, 2015), Nagel (1979), and Strawson (1994, 2018) articulate modern philosophical views of panpsychism, while the physicist Tegmark (2015) speaks of consciousness as a state of matter. Skrbina (2017) provides a readable intellectual history of panpsychist thought. I warmly recommend Teilhard de Chardin (1959). See Fallon (2019b), Morch (2018), and Tononi and Koch (2015) for more on the similarities and differences between IIT and panpsychism.

14. Schrödinger (1958), p. 153.

15. Some variants of dualism are enjoying a resurgence of interest given the failure of physicalism to adequately account for the mental (Owen, 2018b).

16. James (1890), p. 160.

17. The Searle quote is from a criticism of an earlier book of mine with respect to IIT (Searle, 2013a); see note 12 in chapter 13. For a modern formulation of the combination problem, see Goff (2006). Fallon (2019b) discusses in detail how IIT solves the combination problem and addresses Searle's Chinese room argument.

18. Technically, my Whole is neither my entire body nor my entire brain, but only the physical substrate of consciousness within the hot zone in posterior cortex that maximizes *intrinsic* cause-effect power. My body, including my brain, can be considered a maximum of *extrinsic* cause-effect power that breaks down at death.

19. For the architecture of regional sleep, see Koch (2016c) and Vyazonskiy et al. (2011).

20. List (2016) discusses consciousness in group agents, in particular in corporations. He concludes that it does not feel like anything to be a corporation.

21. Such an extension is sometimes referred to as *nonreductive physicalism* (Rowlatt, 2018).

22. Indeed, IIT shares many features with what philosophers refer to as Russellian or neutral monism (Grasso, 2019; Morch, 2018; Russell, 1927).

Coda

1. Several thousand years of philosophical, religious, moral, ethical, scientific, legal, and political perspectives are of relevance here. My only contribution is to make a few noteworthy observations from the point of view of IIT. I found Niikawa (2018) helpful.

2. Human–machine cyborgs bring their own attendant ethical questions pertaining to risks and opportunities for the empowered individual and for society as a whole that will need to be addressed.

3. A tricky question is how to judge the moral standing of nonconscious systems, such as a work of art or an entire ecosystem like a mountain or a natural park. Although they have no intrinsic causal powers as such, they do have attributes worthy of protection.

4. Braithwaite (2010) meticulously documents evidence for pain in fish. The latest estimate for the number of fish annually harvested is at http://fishcount.org.uk/. Tens of billion cattle, pigs, sheep, chickens, ducks, and turkeys are kept under horrendous conditions and are slaughtered each year to feed humanity's relentless appetite for flesh.

5. Computers can be programmed to behave as if they cared. Who has not spent time in customer service hell, being bumped from one voice system to another, each one expressing its great sorrow at having to put us on hold? This sort of fake empathy does not address the existential risks entailed by the creation of superintelligent machines (Bostrom, 2014).

6. Two added complications need to be taken into account. First, the difference between consciousness in a single individual human or animal as measured by its integrated information versus some population measure, say the mean or the median Φ^{max} across a population of adults. Furthermore, any comparison has to account not only for the individual's consciousness at present but its future potential. For instance, a blastocyst in the earliest stage of development, soon after fertilization, with a low Φ^{max} value will eventually develop into a fully conscious adult with a much larger Φ^{max} value.

7. I warmly recommend Singer's *Animal Liberation* (1975) and his *Rethinking Life and Death* (1994).

References

Abe, T., Nakamura, N., Sugishita, M., Kato, Y., & Iwata, M. (1986). Partial disconnection syndrome following penetrating stab wound of the brain. *European Neurology, 25*, 233–239.

Albantakis, L., Hintze, A., Koch, C., Adami, C., & Tononi, G. (2014). Evolution of integrated causal structures in animats exposed to environments of increasing complexity. *PLOS Computational Biology, 10*, e1003966.

Albantakis, L., & Tononi, G. (2015). The intrinsic cause-effect power of discrete dynamical systems—from elementary cellular automata to adapting animats. *Entropy, 17*, 5472–5502.

Almheiri, A., Marolf, D., Polchinski, J., & Sully, J. (2013). Black holes: Complementarity or firewalls? *Journal of High Energy Physics, 2*, 62.

Almog, M., & Korngreen, A. (2016). Is realistic neuronal modeling realistic? *Journal of Neurophysiology, 116*, 2180–2209.

Alonso, L. M., Solovey, G., Yanagawa, T., Proekt, A., Cecchi, G. A., & Magnasco, M. O. (2019). Single-trial classification of awareness state during anesthesia by measuring critical dynamics of global brain activity. *Scientific Reports, 9*, 4927.

Arendt, D., Musser, J. M., Baker, C. V. H., Bergman, A., Cepko, C., Erwin, D. H., Pavlicev, M., Schlosser, G., Widder, S., Laubichler, M. D., & Wagner, G. P. (2016). The origin and evolution of cell types. *Nature Reviews Genetics, 17*, 744–757.

Aru, J., Axmacher, N., Do Lam, A. T., Fell, J., Elger, C. E., Singer, W., & Melloni, L. (2012). Local category-specific gamma band responses in the visual cortex do not reflect conscious perception. *Journal of Neuroscience, 32*, 14909–14914.

Atlan, G., Terem, A., Peretz-Rivlin, N., Sehrawat, K., Gonzales, B. J., Pozner, G., Tasaka, G. I., Goll, Y., Refaeli, R., Zviran, O., & Lim, B. K. (2018). The claustrum supports resilience to distraction. *Current Biology, 28*, 2752–2762.

Azevedo, F., Carvalho, L., Grinberg, L., Farfel, J. M., Ferretti, R., Leite, R., Filho, W. J., Lent, R., & Herculano-Houzel, S. (2009). Equal numbers of neuronal and non-neuronal cells make the human brain an isometrically scaled-up primate brain. *Journal of Comparative Neurology, 513*, 532–541.

Baars, B. J. (1988). *A Cognitive Theory of Consciousness*. Cambridge: Cambridge University Press.

Baars, B. J. (2002). The conscious access hypothesis: Origins and recent evidence. *Trends in Cognitive Sciences, 6*, 47–52.

Bachmann, T., Breitmeyer, B., & Ögmen, H. (2007). *Experimental Phenomena of Consciousness*. New York: Oxford University Press.

Bahney, J., & von Bartheld, C. S. (2018). The cellular composition and glia-neuron ratio in the spinal cord of a human and a nonhuman primate: Comparison with other species and brain regions. *Anatomical Record, 301*, 697–710.

Ball, P. (2019). Neuroscience readies for a showdown over consciousness ideas. *Quanta Magazine,* March 6.

Bardin, J. C., Fins, J. J., Katz, D. I., Hersh, J., Heier, L. A., Tabelow, K., Dyke, J. P., Ballon, D. J., Schiff, N. D., & Voss, H. U. (2011). Dissociations between behavioral and functional magnetic resonance imaging-based evaluations of cognitive function after brain injury. *Brain,* 134, 769–782.

Barrett, A. B. (2016). A comment on Tononi & Koch (2015): Consciousness: Here, there and everywhere? *Philosophical Transactions of the Royal Society B: Biological Sciences, 371*, 20140198.

Barron, A. B., & Klein, C. (2016). What insects can tell us about the origins of consciousness. *Proceedings of the National Academy of Sciences, 113*, 4900–4908.

Barzdin, Y. M. (1993). On the realization of networks in three-dimensional space. In A. N. Shiryayev (Ed.), *Selected Works of A. N. Kolmogorov*. Mathematics and Its Applications (Soviet Series), Vol. 27. Dordrecht: Springer.

Bastuji, H., Frot, M., Perchet, C., Magnin, M., & Garcia-Larrea, L. (2016). Pain networks from the inside: Spatiotemporal analysis of brain responses leading from nociception to conscious perception. *Human Brain Mapping, 37*, 4301–4315.

Bauby, J.-D. (1997). *The Diving Bell and the Butterfly: A Memoir of Life in Death*. New York: Alfred A. Knopf.

Bauer, P. R., Reitsma, J. B., Houweling, B. M., Ferrier, C. H., & Ramsey, N. F. (2014). Can fMRI safely replace the Wada test for preoperative assessment of language lateralization? A meta-analysis and systematic review. *Journal of Neurology, Neurosurgery, and Psychiatry, 85*, 581–588.

Bauman, Z. (2000). *Liquid Modernity*. Cambridge: Polity.

Beauchamp, M. S., Sun, P., Baum, S. H., Tolias, A. S., & Yoshor, D. (2013). Electrocorticography links human temporoparietal junction to visual perception. *Nature Neuroscience, 15*, 957–959.

Beilock, S. L., Carr, T. H., MacMahon, C., & Starkes, J. L. (2002). When paying attention becomes counterproductive: Impact of divided versus skill-focused attention on novice and experienced performance of sensorimotor skills. *Journal of Experimental Psychology: Applied, 8*, 6–16.

Bellesi, M., Riedner, B.A,, Garcia-Molina, G.N., Cirelli, C., & Tononi, G. (2014). Enhancement of sleep slow waves: underlying mechanisms and practical consequences. *Frontiers in Systems Neuroscience, 8*, 208–218.

Bengio, Y. (2017). The consciousness prior. *arXiv*, 1709.08568v1.

Berlin, H. A. (2011). The neural basis of the dynamic unconscious. *Neuropsychoanalysis, 13*, 5–31.

Berlin, H. A., & Koch, C. (2009). Neuroscience meets psychoanalysis. *Scientific American Mind*, April, 16–19.

Biderman, N., & Mudrik, L. (2017). Evidence for implicit—but not unconscious—processing of object-scene relations. *Psychological Science, 29*, 266–277.

Birey, F., Andersen, J., Makinson, C. D., Islam, S., Wei, W., Huber, N., Fan, H. C., Metzler, K. R. C., Panagiotakos, G., Thom, N., & O'Rourke, N. A. (2017). Assembly of functionally integrated human forebrain spheroids. *Nature, 545*, 54–59.

Blackmore, S. (2011). *Zen and the Art of Consciousness*. London: One World Publications.

Blanke, O., Landis, T., & Seeck M. (2000). Electrical cortical stimulation of the human prefrontal cortex evokes complex visual hallucinations. *Epilepsy & Behavior, 1*, 356–361.

Block, N. (1995). On a confusion about a function of consciousness. *Behavioral and Brain Sciences, 18*, 227–287.

Block, N. (2007). Consciousness, accessibility, and the mesh between psychology and neuroscience. *Behavioral and Brain Sciences, 30*, 481–548.

Block, N. (2011). Perceptual consciousness overflows cognitive access. *Trends in Cognitive Sciences, 15*, 567–575.

Bogen, J. E. (1993). The callosal syndromes. In K. M. Heilman & E. Valenstein (Eds.), *Clinical Neurosychology* (3rd ed., pp. 337–407). New York: Oxford University Press.

Bogen, J. E., & Gordon, H. W. (1970). Musical tests for functional lateralization with intracarotid amobarbital. *Nature, 230*, 524–525.

Bostrom, N. (2014). *Superintelligence: Paths, Dangers, Strategies*. Oxford: Oxford University Press.

Boyd, C. A. (2010). Cerebellar agenesis revisited. *Brain, 133*, 941–944.

Bolte Taylor, J. (2008). *My Stroke of Insight: A Brain Scientist's Personal Journey*. New York: Viking.

Boly, M., Massimini, M., Tsuchiya, N., Postle, B. R., Koch, C., & Tononi, G. (2017). Are the neural correlates of consciousness in the front or in the back of the cerebral cortex? Clinical and neuroimaging evidence. *Journal of Neuroscience, 37*, 9603–9613.

Bornhövd, K., Quante, M., Glauche, V., Bromm, B., Weiller, C., & Büchel, C. (2002). Painful stimuli evoke different stimulus-response functions in the amygdala, prefrontal, insula and somatosensory cortex: A single-trial fMRI study. *Brain, 125*, 1326–1336.

Braitenberg, V., & Schüz, A. (1998). *Cortex: Statistics and Geometry of Neuronal Connectivity* (2nd ed.). Berlin: Springer.

Braithwaite, V. (2010). *Do Fish Feel Pain?* Oxford: Oxford University Press.

Braun, J., & Julesz, B. (1998). Withdrawing attention at little or no cost: Detection and discrimination tasks. *Perception & Psychophysics, 60*, 1–23.

Brickner, R. M. (1936). *The Intellectual Functions of the Frontal Lobes*. New York: MacMillan.

Brickner, R. M. (1952). Brain of Patient A. after bilateral frontal lobectomy: Status of frontal-lobe problem. *AMA Archives of Neurology and Psychiatry, 68*, 293–313.

Bruner, J. C., & Potter, M. C. (1964). Interference in visual recognition. *Science, 114*, 424–425.

Bruno, M.-A., Nizzi, M.-C., Laureys, S., & Gosseries, O. (2016). Consciousness in the locked-in syndrome. In Laureys, S., Gosseries, O., & Tononi, G. (Eds.), *The Neurology of Consciousness* (2nd ed., pp. 187–202). Amsterdam: Elsevier.

Button, K. S., Ioannidis, J. P.A., Mokrysz, C., Nosek, B. A., Flint, J., Robinson, E. S., & Munafo, M. R. (2013). Power failure: Why small sample size undermines the reliability of neuroscience. *Nature Reviews Neuroscience, 14*, 365–376.

Buzsaki, G., Anastassiou, C. A., & Koch, C. (2012). The origin of extracellular fields and currents—EEG, ECoG, LFP and spikes. *Nature Reviews Neuroscience, 13*, 407–420.

Carlen, M. (2017). What constitutes the prefrontal cortex? *Science, 358*, 478–482.

Carmel, D., Lavie, N., & Rees, G. (2006). Conscious awareness of flicker in humans involves frontal and parietal cortex. *Current Biology, 16*, 907–911.

Casali, A., Gosseries, O., Rosanova, M., Boly, M., Sarasso, S., Casali, K. R., Casarotto, S., Bruno, M. A., Laureys, S., Tononi, G., & Massimini, M. (2013). A theoretically based index of consciousness independent of sensory processing and behavior. *Science Translational Medicine, 5,* 1–11.

Casarotto, S., Comanducci, A., Rosanova, M., Sarasso, S., Fecchio, M., Napolitani, M., et al. (2016). Stratification of unresponsive patients by an independently validated index of brain complexity. *Annals of Neurology, 80,* 718–729.

Casti, J. L., & DePauli, W. (2000). *Gödel: A Life of Logic.* New York: Basic Books.

Cauller, L. J., & Kulics, A. T. (1991). The neural basis of the behaviorally relevant N1 component of the somatosensory-evoked potential in SI cortex of awake monkeys: Evidence that backward cortical projections signal conscious touch sensation. *Experimental Brain Research, 84,* 607–619.

Cerf, M., Thiruvengadam, N., Mormann, F., Kraskov, A., Quian Quiroga, R., Koch, C., & Fried, I. (2010). On-line, voluntary control of human temporal lobe neurons. *Nature, 467,* 1104-1108.

Chalmers, D. J. (1996). *The Conscious Mind: In Search of a Fundamental Theory.* New York: Oxford University Press.

Chalmers, D. J. (1998). On the search for the neural correlate of consciousness. In S. Hameroff, A. Kaszniak, & A. Scott (Eds.), *Toward a Science of Consciousness II: The Second Tucson Discussions and Debates.* Cambridge, MA: MIT Press.

Chalmers, D. J. (2000). What is a neural correlate of consciousness? In T. Metzinger (Ed.), *Neural Correlates of Consciousness: Empirical and Conceptual Questions* (pp. 17–39). Cambridge, MA: MIT Press.

Chalmers, D. J. (2015) Panpsychism and panprotopsychism. In T. Alter & Y. Nagasawa (Eds.), *Consciousness in the Physical World: Perspectives on Russellian Monism* (pp. 246–276). New York: Oxford University Press.

Chamovitz, D. (2012). *What a Plant Knows: A Field Guide to the Sense.* New York: Scientific American/Farrar, Straus and Giroux.

Chiao, R. Y., Cohen, M. L., Leggett, A. J., Phillips, W. D., & Harper Jr., C. L. (Eds.). (2010). *Amazing Light: New Light on Physics, Cosmology and Consciousness.* Cambridge: Cambridge University Press.

Churchland, Patricia. (1983). Consciousness: The transmutation of a concept. *Pacific Philosophical Quarterly, 64,* 80–95.

Churchland, Patricia. (1986). *Neurophilosophy—Toward a Unified Science of the Mind/Brain.* Cambridge, MA: MIT Press.

Churchland, Paul. (1984). *Matter and Consciousness: A Contemporary Introduction to the Philosophy of Mind.* Cambridge, MA: MIT Press.

Cignetti, F., Vaugoyeau, M., Nazarian, B., Roth, M., Anton, J. L., & Assaiante, C. (2014). Boosted activation of right inferior frontoparietal network: A basis for illusory movement awareness. *Human Brain Mapping, 35*, 5166–5178.

Clark, A. (1989). *Microcognition: Philosophy, Cognitive Science, and Parallel Distributed Processing*. Cambridge, MA: MIT Press.

Clarke, A. (1962). *Profiles of the Future: An Inquiry into the Limits of the Possible*. New York: Bantam Books.

Clarke, A. (1963). Aristotelian concepts of the form and function of the brain. *Bulletin of the History of Medicine, 37*, 1–14.

Cohen, M. A., & Dennett, D. C. (2011). Consciousness cannot be separated from function. *Trends in Cognitive Sciences, 15*, 358–364.

Cohen, M. A., Dennett, D. C., & Kanwisher, N. (2016). What is the bandwidth of perceptual experience? *Trends in Cognitive Sciences, 20*, 324–335.

Collini, E., Wong, C. Y., Wilk, K. E., Curmi, P. M. G., Brumer, P., & Schoes, G. D. (2010). Coherently wired light-harvesting in photosynthetic marine algae at ambient temperature. *Nature, 463*, 644–647.

Comolatti, R., Pigorini, A., Casarotto, S., Fecchio, M., Faria, G., Sarasso, S., Rosanova, M., Gosseries, O., Boly, M., Bodart, O., Ledou, D., Brichant, J. F., Nobili, L., Laureys, S., Tononi, G., Massimini, M., & Casali, A. G. (2018). A fast and general method to empirically estimate the complexity of distributed causal interactions in the brain. *bioRxiv*. doi:10.1101/445882.

Cook, M. (2004). Universality in elementary cellular automata. *Complex Systems, 15*, 1–40.

Cottingham, J. (1978). A brute to the brutes—Descartes' treatment of animals. *Philosophy, 53*, 551–559.

Cranford, R. (2005). Facts, lies, and videotapes: The permanent vegetative state and the sad case of Terri Schiavo. *Journal of Law, Medicine & Ethics, 33*, 363–371.

Crick, F. C. (1988). *What Mad Pursuit*. New York: Basic Books.

Crick, F. C. (1994). *The Astonishing Hypothesis*. New York: Charles Scribner's Sons.

Crick, F. C., & Koch, C. (1990). Towards a neurobiological theory of consciousness. *Seminars in Neuroscience, 2*, 263–275.

Crick, F. C., & Koch, C. (1995). Are we aware of neural activity in primary visual cortex? *Nature, 375*, 121–123.

Crick, F. C., & Koch, C. (1998). Consciousness and neuroscience. *Cerebral Cortex, 8*, 97–107.

Crick, F. C., & Koch, C. (2000). The unconscious homunculus. With commentaries by multiple authors. *Neuro-Psychoanalysis, 2*, 3–59.

Crick, F. C., & Koch, C. (2005). What is the function of the claustrum? *Philosophical Transactions of the Royal Society of London B: Biological Sciences, 360*, 1271–1279.

Curtiss, S. (1977). *Genie: A Psycholinguistic Study of a Modern-Day "Wild Child."* Perspectives in Neurolinguistics and Psycholinguistics. Boston: Academic Press.

Cybenko, G. (1989). Approximations by superpositions of sigmoidal functions. *Mathematics of Control, Signals, and Systems, 2*, 303–314.

Dakpo Tashi Namgyal (2004). *Clarifying the Natural State*. Hong Kong: Rangjung Yeshe Publications.

Darwin, C. (1881/1985). *The Formation of Vegetable Mould, through the Action of Worms with Observation of their Habits*. Chicago: University of Chicago Press.

Dawid, R. (2013). *String Theory and the Scientific Method*. Cambridge: Cambridge University Press.

Dawkins, M. S. (1998). *Through Our Eyes Only—The Search for Animal Consciousness*. Oxford: Oxford University Press.

Dean, P., Porrill, J., Ekerot, C. F., & Jörntell, H. (2010). The cerebellar microcircuit as an adaptive filter: Experimental and computational evidence. *Nature Reviews Neuroscience, 11*, 30–43.

Deary, I. J., Penke, L., & Johnson, W. (2010). The neuroscience of human intelligence differences. *Nature Reviews Neuroscience, 11*, 201–211.

Debellemaniere, E., Chambon, S., Pinaud, C., Thorey, V., Dehaene, D., Léger, D., Mounir, C., Arnal, P. J., & Galtier, M. N. (2018). Performance of an ambulatory dry-EEG device for auditory closed-loop stimulation of sleep slow oscillations in the home environment. *Frontiers in Human Neuroscience, 12*, 88.

Dehaene, S. (2014). *Consciousness and the Brain: Deciphering How the Brain Codes Our Thoughts*. New York: Viking.

Dehaene, S., & Changeux, J.-P. (2011). Experimental and theoretical approaches to conscious processing. *Neuron, 70*, 200–227.

Dehaene, S., Changeux, J.-P., Naccache, L., Sackur, J., & Sergent, C. (2006). Conscious, preconscious, and subliminal processing: A testable taxonomy. *Trends in Cognitive Sciences, 10*, 204–211.

Dehaene, S., Lau, H., & Kouider, S. (2017). What is consciousness, and could machines have it? *Science, 358*, 486–492.

Dehaene, S., Naccache, L., Cohen, L., Le Bihan, D., Mangin, J.-F., Poline, J.-B., et al. (2001). Cerebral mechanisms of word masking and unconscious repetition priming. *Nature Neuroscience, 4*, 752–758.

de Lafuente, V., & Romo, R. (2006). Neural correlate of subjective sensory experience gradually builds up across cortical areas. *Proceedings of the National Academy of Sciences, 103*, 14266–14271.

Del Cul, A., Dehaene, S., Reyes, P., Bravo, E., & Slachevsky, A. (2009). Causal role of prefrontal cortex in the threshold for access to consciousness. *Brain, 132*, 2531–2540.

Dement, W. C., & Vaughan, C. (1999). *The Promise of Sleep*. New York: Dell.

Dennett, D. C. (1991). *Consciousness Explained*. Boston: Little, Brown.

Dennett, D. C. (2017). *From Bacteria to Bach and Back: The Evolution of Minds*. New York: W. W. Norton.

Denton, D. (2006). *The Primordial Emotions: The Dawning of Consciousness*. Oxford: Oxford University Press.

Desmurget, M., Reilly, K. T., Richard, N., Szathmari, A., Mottolese, C., & Sirigu, A. (2009). Movement intention after parietal cortex stimulation in humans. *Science, 324*, 811–813.

Desmurget, M., Song, Z., Mottolese, C., & Sirigu, A. (2013). Re-establishing the merits of electrical brain stimulation. *Trends in Cognitive Sciences, 17*, 442–449.

de Vries, S. E., Lecoq, J., Buice, M. A., Groblewski, P. A., Ocker, G. K., Oliver, M. et al. (2018). A large-scale, standardized physiological survey reveals higher order coding throughout the mouse visual cortex. *bioRxiv*, 359513.

Di Lullo, E., & Kriegstein, A. R. (2017). The use of brain organoids to investigate neural development and disease. *Nature Reviews Neuroscience, 1*, 573–583.

Dominus, S. (2011). Could conjoined twins share a mind? *New York Times Magazine*, May 25.

Dougherty, R. F., Koch, V. M., Brewer, A. A., Fischer, B., Modersitzki, J., & Wandell, B. A. (2003). Visual field representations and locations of visual areas V1/2/3 in human visual cortex. *Journal of Vision, 3*, 586–598.

Doyen, S., Klein, P., Lichon, C.-L., & Cleeremans, A. (2012). Behavioral priming: It's all in the mind, but whose mind? *PLOS One, 7*. doi:10.1371/journal.pone.0029081.

Drews, F. A., Pasupathi, M., & Strayer, D. L. (2008). Passenger and cell phone conversations in simulated driving. *Journal of Experimental Psychology Applied, 14*, 392–400.

Earl, B. (2014). The biological function of consciousness. *Frontiers in Psychology, 5*, 697.

Economo, M. N., Clack, N. G., Lavis, L. D., Gerfen, C. R., Svoboda, K., Myers, E. W., & Chandrashekar, J. (2016). A platform for brain-wide imaging and reconstruction of individual neurons. *eLife, 5*, 10566.

Edelman, G. M., & Tononi, G. (2000). *A Universe of Consciousness*. New York: Basic Books.

Edlund, J. A., Chaumont, N., Hintze, A., Koch, C., Tononi, G., & Adami, C. (2011). Integrated information increases with fitness in the evolution of animats. *PLOS Computational Biology, 7*, e1002236.

Engel, A. K., & Singer, W. (2001). Temporal binding and the neural correlates of sensory awareness. *Trends in Cognitive Sciences, 5*, 16–25.

Eslami, S. M. A., Rezende, D. J., Desse, F., Viola, F., Morcos, A. S., Garnelo, M., et al. (2018). Neural representation and rendering. *Science, 360*, 1204–1210.

Esteban, F. J., Galadi, J., Langa, J. A., Portillo, J. R., & Soler-Toscano, F. (2018). Informational structures: A dynamical system approach for integrated information. *PLOS Computational Biology, 14*. doi:10.1371/journal.pcbi.1006154.

Everett, D. L. (2008). *Don't Sleep, There Are Snakes—Life and Language in the Amazonian Jungle*. New York: Vintage.

Fallon, F. (2019a). Dennett on consciousness: Realism without the hysterics. *Topoi*, in press. doi:10.1007/s11245-017-9502-8.

Fallon, F. (2019b). Integrated information theory, Searle, and the arbitrariness question. *Review of Philosophy and Psychology*, in press.

Farah, M. J. (1990). *Visual Agnosia*. Cambridge, MA: MIT Press.

Fei-Fei, L., Iyer, A., Koch, C., & Perona, P. (2007). What do we perceive in a glance of a real-world scene? *Journal of Vision, 7*, 1–29.

Fei-Fei, L., VanRullen, R., Koch, C., & Perona, P. (2005). Why does natural scene categorization require little attention? Exploring attentional requirements for natural and synthetic stimuli. *Visual Cognition, 12*, 893–924.

Feinberg, T. E., & Mallatt, J. M. (2016). *The Ancient Origins of Consciousness*. Cambridge, MA: MIT Press.

Feinberg, T. E., Schindler, R. J., Flanagan, N. G., & Haber, L. D. (1992). Two alien hand syndromes. *Neurology, 42*, 19–24.

Findlay, G., Marshall, W., Albantakis, L., Mayner, W., Koch, C., & Tononi, G. (2019). Can computers be conscious? Dissociating functional and phenomenal equivalence. Submitted.

Flaherty, M. G. (1999). *A Watched Pot: How We Experience Time*. New York: NYU Press.

Fleming, S. M., Ryu, J., Golfinos, J. G., & Blackmon, K. E. (2014). Domain-specific impairment in metacognitive accuracy following anterior prefrontal lesions. *Brain, 137*, 2811–2822.

Forman, R. K. C. (1990a). Eckhart, *Gezücken*, and the ground of the soul. In R. K. C. Forman (Ed.), *The Problem of Pure Consciousness* (pp. 98–120). New York: Oxford University Press.

Forman, R. K. C. (Ed.). (1990b). *The Problem of Pure Consciousness*. New York: Oxford University Press.

Forman, R. K. C. (1990c). Introduction: Mysticism, constructivism, and forgetting. In R. K. C. Forman (Ed.), *The Problem of Pure Consciousness* (pp. 3–49). New York: Oxford University Press.

Foster, B. L., & Parvizi, J. (2107). Direct cortical stimulation of human posteromedial cortex. *Neurology, 88*, 1–7.

Fox, K. C., Yih, J., Raccah, O., Pendekanti, S. L., Limbach, L. E., Maydan, D. D., & Parvizi, J. (2018). Changes in subjective experience elicited by direct stimulation of the human orbitofrontal cortex. *Neurology, 91*, e1519–e1527.

Freud, S. (1915). The unconscious. In *The Standard Edition of the Complete Psychological Works of Sigmund Freud*, 14:159–204. London: Hogarth Press.

Freud, S. (1923). The ego and the id. In *The Standard Edition of the Complete Psychological Works of Sigmund Freud*, 19:1–59. London: Hogarth Press.

Friedmann, S., Schemmel, J., Grübl, A., Hartel, A., Hock, M., & Meier, K. (2017). Demonstrating hybrid learning in a flexible neuromorphic hardware system. *IEEE Transactions on Biomedical Circuits and Systems, 11*, 128–142.

Fries, P., Roelfsema, P. R., Engel, A. K., König, P., & Singer, W. (1997). Synchronization of oscillatory responses in visual cortex correlates with perception in interocular rivalry. *Proceedings of the National Academy of Sciences, 94*, 12699–12704.

Friston, K. (2010). The free-energy principle: A unified brain theory? *Nature Reviews Neurosciences, 11*, 127–138.

Gagliano, M. (2017). The mind of plants: Thinking the unthinkable. *Communicative & Integrative Biology, 10*, e1288333.

Gagliano, M., Vyazovskiy, V. V., Borbély, A. A., Mavra Grimonprez, M., & Depczynski, M. (2016). Learning by association in plants. *Scientific Reports, 6*, 38427.

Gallant, J. L., Shoup, R. E., & Mazer J. A. (2000). A human extrastriate area functionally homologous to macaque V4. *Neuron, 27*, 227–235.

Gauthier, I. (2017). The quest for the FFA led to the expertise account of its specialization. *arXiv*, 1702.07038.

Genetti, M., Britz, J., Michel, C. M., & Pegna, A. J. (2010). An electrophysiological study of conscious visual perception using progressively degraded stimuli. *Journal of Vision, 10,* 1–14.

Giacino, J. T., Fins, J. J., Laureys, S., & Schiff, N. D. (2014). Disorders of consciousness after acquired brain injury: The state of the science. *Nature Reviews Neuroscience, 10,* 99–114.

Giattino, C. M., Alam, Z. M., & Woldorff, M. G. (2017). Neural processes underlying the orienting of attention without awareness. *Cortex, 102,* 14–25.

Giurfa, M., Zhang, S., Jenett, A., Menzel, R., & Srinivasan, M. V. (2001). The concepts of "sameness" and "difference" in an insect. *Nature, 410,* 930–933.

Godfrey-Smith, P. (2016). *Other Minds—The Octopus, the Sea and the Deep Origins of Consciousness.* New York: Farrar, Straus & Giroux.

Goff, P. (2006). Experiences don't sum. *Journal of Consciousness Studies, 13,* 53–61.

Gordon, H. W., & Bogen, J. E. (1974). Hemispheric lateralization of singing after intracarotid sodium amylobarbitone. *Journal of Neurology, Neurosurgery & Psychiatry, 37,* 727–738.

Goriounova, N. A., Heyer, D. B., Wilbers, R., Verhoog, M. B., Giugliano, M., Verbist, C., et al. (2019). A cellular basis of human intelligence. *eLife,* in press.

Grasso, M. (2019). IIT vs. Russellian Monism: A metaphysical showdown on the content of experience. *Journal of Consciousness Studies, 26,* 48–75.

Gratiy, S., Geir, H., Denman, D., Hawrylycz, M., Koch, C., Einevoll, G. & Anastassiou, C. (2017). From Maxwell's equations to the theory of current-source density analysis. *European Journal of Neuroscience, 45,* 1013–1023.

Greenwood, V. (2015). Consciousness began when gods stopped speaking. *Nautilus,* May 28.

Griffin, D. R. (1998). *Unsnarling the World-Knot—Consciousness, Freedom and the Mind-Body problem.* Eugene, OR: Wipf & Stock.

Griffin, D. R. (2001). *Animal Minds—Beyond Cognition to Consciousness.* Chicago: University of Chicago Press.

Gross, G. G. (1998). *Brain, Vision, Memory—Tales in the History of Neuroscience.* Cambridge, MA: MIT Press.

Hadamard, J. (1945). *The Mathematician's Mind.* Princeton: Princeton University Press.

Hameroff, S., & Penrose, R. (2014). Consciousness in the universe: A review of the "Orch OR" theory. *Physics Life Reviews, 11,* 39–78.

Han, E., Alais, D., & Blake, R. (2018). Battle of the Mondrians: Investigating the role of unpredictability in continuous flash suppression. *I-Perception, 9,* 1–21.

Harris, C. R., Coburn, N., Rohrer, D., & Pashler, H. (2013). Two failures to replicate high-performance-goal priming effects. *PLOS One, 8,* e72467.

Harrison, P. (2015). *The Territories of Science and Religion.* Chicago: University of Chicago Press.

Harth, E. (1993). *The Creative Loop: How the Brain Makes a Mind.* Reading, MA: Addison-Wesley.

Hasenkamp, W. (Ed.). (2017). *The Monastery and the Microscope—Conversations with the Dalai Lama on Mind, Mindfulness and the Nature of Reality.* New Haven: Yale University Press.

Hassenfeld, S. (2018). *Lost in Mathematics: How Beauty Leads Physics Astray.* New York: Basic Books.

Hassin, R. R., Uleman, J. S., & Bargh, J. A. (2005). *The New Unconscious.* Oxford: Oxford University Press.

Haun, A. M., Tononi, G., Koch, C., & Tsuchiya, N. (2017). Are we underestimating the richness of visual experience? *Neuroscience of Consciousness, 1,* 1–4.

Haynes, J. D., & Rees, G. (2005). Predicting the orientation of invisible stimuli from activity in human primary visual cortex. *Nature Neuroscience, 8,* 686–691.

Hebb, D. O., & Penfield, W. (1940). Human behavior after extensive bilateral removal from the frontal lobes. *Archives of Neurology and Psychiatry, 42,* 421–438.

Henri-Bhargava, A., Stuff, D.T., & Freedman, M. (2018). Clinical assessment of prefrontal lobe functions. *Behavioral Neurology and Psychiatry, 24,* 704–726.

Hepp, K. (2018). The wake-sleep "phase transition" at the gate to consciousness. *Journal of Statistical Physics, 172,* 562–568.

Herculano-Houzel, S., Mota, B., & Lent, R. (2006). Cellular scaling rules for rodent brains. *Proceedings of the National Academy of Sciences, 103,* 12138–12143.

Herculano-Houzel, S., Munk, M. H., Neuenschwander, S., & Singer, W. (1999). Precisely synchronized oscillatory firing patterns require electroencephalographic activation. *Journal of Neuroscience, 19,* 3992–4010.

Hermes, D., Miller, K. J., Wandell, B. A., & Winawer, J. (2015). Stimulus dependence of gamma oscillations in human visual cortex. *Cerebral Cortex, 25,* 2951–2959.

Heywood, C. A., & Zihl, J. (1999). Motion blindness. In G. W. Humphreys (Ed.), *Case Studies in the Neuropsychology of Vision* (pp. 1–16). Hove: Psychology Press/ Taylor & Francis.

Hiscock, H. G., Worster, S., Kattnig, D. R., Steers, C., Jin, Y., Manolopoulos, D. E., Mouritsen, H., & Hore, P. J. (2016). The quantum needle of the avian magnetic compass. *Proceedings of the National Academy of Sciences, 113*, 4634–4639.

Hodgkin, A. L. (1976). Chance and design in electrophysiology: An informal account of certain experiments on nerve carried out between 1934 and 1952. *Journal of Physiology, 263*, 1–21.

Hodgkin, A. L., & Huxley, A. F. (1952). A quantitative description of membrane current and its application to conduction and excitation in nerve. *Journal of Physiology, 117*, 500–544.

Hoel, E. P., Albantakis, L., & Tononi, G. (2013). Quantifying causal emergence shows that macro can beat micro. *Proceedings of the National Academy of Sciences, 110*, 19790–19795.

Hohwy, J. (2013). *The Predictive Mind.* Oxford: Oxford University Press.

Holsti, L., Grunau, R. E., & Shany, E. (2011). Assessing pain in preterm infants in the neonatal intensive care unit: Moving to a "brain-oriented" approach. *Pain Management, 1*, 171–179.

Holt, J. (2012). *Why Does the World Exist?* New York: W. W. Norton.

Hornik, K., Stinchcombe, M., & White, H. (1989). Multilayer feedforward networks are universal approximators. *Neural Networks, 2*, 359–366.

Horschler, D. J., Hare, B., Call, J., Kaminski, J., Miklosi, A., & MacLean, E. L. (2019). Absolute brain size predicts dog breed differences in executive function. *Animal Cognition, 22*, 187–198.

Hsieh, P. J., Colas, J. T., & Kanwisher, N. (2011). Pop-out without awareness: Unseen feature singletons capture attention only when top-down attention is available. *Psychological Science, 22*, 1220–1226.

Hubel, D. H. (1988). *Eye, Brain, and Vision.* New York: Scientific American Library.

Hyman, I. E., Boss, S. M., Wise, B. M., McKenzie, K. E., & Caggiano, J. M. (2010). Did you see the unicycling clown? Inattentional blindness while walking and talking on a cell phone. *Applied Cognitive Psychology, 24*, 597–607.

Imas, O. A., Ropella, K. M., Ward, B. D., Wood, J. D., & Hudetz, A. G. (2005). Volatile anesthetics disrupt frontal-posterior recurrent information transfer at gamma frequencies in rat. *Neuroscience Letters, 387*, 145–150.

Ioannidis, J. P. A. (2017). Are most published results in psychology false? An empirical study. https://replicationindex.wordpress.com/2017/01/15/are-most-published-results-in-psychology-false-an-empirical-study/.

Irvine, E. (2013). *Consciousness as a Scientific Concept: A Philosophy of Science Perspective*. Springer: Heidelberg.

Itti, L., & Baldi, P. (2006). Bayesian surprise attracts human attention: Advances in neural information processing systems. In *NIPS 2005* (Vol. 19, pp. 547–554). Cambridge, MA: MIT Press.

Itti, L., & Baldi, P. (2009). Bayesian surprise attracts human attention. *Vision Research, 49*, 1295–1306.

Ius, T., Angelini, E., Thiebaut de Schotten, M., Mandonnet, E., & Duffau, H. (2011). Evidence for potentials and limitations of brain plasticity using an atlas of functional resectability of WHO grade II gliomas: Towards a "minimal common brain." *NeuroImage, 56*, 992–1000.

Jackendoff, R. (1987). *Consciousness and the Computational Mind*. Cambridge, MA: MIT Press.

Jackendoff, R. (1996). How language helps us think. *Pragmatics & Cognition, 4*, 1–34.

Jackson, J., Karnani, M. M., Zemelman, B. V., Burdakov, D., & Lee, A. K. (2018). Inhibitory control of prefrontal cortex by the claustrum. *Neuron, 99*, 1029–1039.

Jakob, J., Tammo, I., Moritz, H., Itaru, K., Mitsuhisa, S., Jun, I., Markus, D., & Susanne, K. (2018). Extremely scalable spiking neuronal network simulation code: From laptops to exascale computers. *Frontiers in Neuroscience, 12*. doi:10.3389/fninf.2018.00002.

James, W. (1890). *The Principles of Psychology*. New York: Holt.

Jaynes, J. (1976). *The Origin of Consciousness in the Breakdown of the Bicameral Mind*. Boston: Houghton Mifflin.

Jennings, H. S. (1906). *Behavior of the Lower Organisms*. New York: Columbia University Press.

Jiang, Y., Costello, P., Fang, F., Huang, M., & He, S. (2006). A gender- and sexual orientation-dependent spatial attentional effect of invisible images. *Proceedings of the National Academy of Sciences, 103*, 17048–17052.

Johansson, P., Hall, L., Sikström, S., & Olsson, A. (2005). Failure to detect mismatches between intention and outcome in a simple decision task. *Science, 310*, 116–119.

Johnson-Laird, P. N. (1983). A computational analysis of consciousness. *Cognition & Brain Theory, 6*, 499–508.

Jordan, G., Deeb, S. S., Bosten, J. M., & Mollon, J. D. (2010). The dimensionality of color vision carriers of anomalous trichromacy. *Journal of Vision, 10*, 12.

Jordan, J., Ippen, T., Helias, M., Kitayama, I., Sato, M., Igarashi, J., Diesmann, M. D., & Kunkel, S. (2018). Extremely scalable spiking neuronal network simulation code:

From laptops to exascale computers. *Frontiers in Neuroinformatics, 12*. doi:10.3389/fninf.2018.00002.

Joshi, N. J., Tononi, G., & Koch, C. (2013). The minimal complexity of adapting agents increases with fitness. *PLOS Computational Biology, 9*, e1003111.

Jun, J. J., Steinmetz, N. A., Siegle, J. H., Denman, D. J., Bauza, M., Barbarits, B., et al. (2017). Fully integrated silicon probes for high density recording of neural activity. *Nature, 551*, 232–236.

Kaas, J. H. (2009). The evolution of sensory and motor systems in primates. In J. H. Kaas (Ed.), *Evolutionary Neuroscience* (pp. 523–544). New York: Academic Press.

Kannape, O. A., Perrig, S., Rossetti, A. O., & Blanke, O. (2017). Distinct locomotor control and awareness in awake sleepwalkers. *Current Biology, 27*, R11-2-1104.

Kanwisher, N. (2017). The quest for the FFA and where it led. *Journal of Neuroscience, 37*, 1056–1061.

Kanwisher, N., McDermott, J., & Chun, M. M. (1997). The fusiform face area: a module in human extrastriate cortex specialized for face perception. *Journal of Neuroscience, 17*, 4302–4311.

Karinthy, F. (1939). *A Journey Round My Skull*. New York: New York Review of Books.

Karten, H. J. (2015). Vertebrate brains and the evolutionary connectomics: On the origins of the mammalian neocortex. *Philosophical Transactions of the Royal Society of London B: Biological Sciences, 370*, 20150060.

Karr, J. R., Sanghvi, J. C., Macklin, D. N., Gutschow, M. V., Jacobs, J. M., Bolival, B. Jr., et al. (2012). A whole-cell computational model predicts phenotype from genotype. *Cell, 150*, 389–401.

Keefe, P. R. (2016). The detectives who never forget a face. *New Yorker*, August 22.

Kertai, M. D., Whitlock, E. L., & Avidan, M. S. (2012). Brain monitoring with electroencephalography and the electroencephalogram-derived bispectral index during cardiac surgery. *Anesthesia & Analgesia, 114*, 533–546.

Killingsworth, M. A., & Gilbert, D. T. (2010). A wandering mind is an unhappy mind. *Science, 330*, 932.

King, B. J. (2013). *How Animals Grieve*. Chicago: University of Chicago Press.

King, J. R., Sitt, J. D., Faugeras, F., Rohaut, B., El Karoui, I., Cohen, L., et al. (2013). Information sharing in the brain indexes consciousness in noncommunicative patients. *Current Biology, 23*, 1914–1919.

Koch, C. (1999). *Biophysics of Computation: Information Processing in Single Neurons*. New York: Oxford University Press.

Koch, C. (2004). *The Quest for Consciousness: A Neurobiological Approach.* Denver: Roberts.

Koch, C. (2012a). *Consciousness: Confessions of a Romantic Reductionist.* Cambridge, MA: MIT Press.

Koch, C. (2012b). Modular biological complexity. *Science, 337,* 531–532.

Koch, C. (2014). A brain structure looking for a function. *Scientific American Mind,* November, 24–27.

Koch, C. (2015). Without a thought. *Scientific American Mind,* May, 25–26.

Koch, C. (2016a). Does brain size matter? *Scientific American Mind,* January, 22–25.

Koch, C. (2016b). Sleep without end. *Scientific American Mind,* March, 22–25.

Koch, C. (2016c). Sleeping while awake. *Scientific American Mind,* November, 20–23.

Koch, C. (2017a). Contacting stranded minds. *Scientific American Mind,* May, 20–23.

Koch, C. (2017b). The feeling of being a brain: Material correlates of consciousness. In W. Hasenkamp (Ed.), *The Monastery and the Microscope—Conversations with the Dalai Lama on Mind, Mindfulness and the Nature of Reality* (pp. 112–141). New Haven: Yale University Press.

Koch, C. (2017c). God, death and Francis Crick. In C. Burns (Ed.), *All These Wonders: True Stories Facing the Unknown* (pp. 41–50). New York: Crown Archetype.

Koch, C. (2017d). How to make a consciousness meter. *Scientific American,* November, 28–33.

Koch, C., & Buice, M. A. (2015). A biological imitation game. *Cell, 163,* 277–280.

Koch, C., & Crick, F. C. (2001). On the zombie within. *Nature, 411,* 893.

Koch, C., & Hepp, K. (2006). Quantum mechanics and higher brain functions: Lessons from quantum computation and neurobiology. *Nature, 440,* 611–612.

Koch, C. & Hepp, K. (2010). The relation between quantum mechanics and higher brain functions: Lessons from quantum computation and neurobiology. In R. Y. Chiao et al. (Eds.), *Amazing Light: New Light on Physics, Cosmology and Consciousness* (pp. 584–600). Cambridge: Cambridge University Press.

Koch, C., & Jones, A. (2016). Big science, team science, and open science for neuroscience. *Neuron, 92,* 612–616.

Koch, C., Massimini, M., Boly, M., & Tononi, G. (2016). The neural correlates of consciousness: Progress and problems. *Nature Review Neuroscience, 17,* 307–321.

Koch, C., McLean, J., Segev, R., Freed, M. A., Berryll, M. J., Balasubramanian, V., & Sterling, P. (2006). How much the eye tells the brain. *Current Biology, 16,* 1428–1434.

Koch, C., & Segev, I. (Eds.). (1998). *Methods in Neuronal Modeling: From Ions to Networks*. Cambridge, MA: MIT Press.

Koch, C., & Tononi, G. (2013). Letter to the Editor: Can a photodiode be conscious? *New York Review of Books, 60*, 43.

Kohler, C. G., Ances, B. M., Coleman, A. R., Ragland, J. D., Lazarev, M., & Gur, R. C. (2000). Marchiafava-Bignami disease: literature review and case report. *Neuropsychiatry Neuropsychology and Behavioral Neurology, 13*(1), 67–67.

Kotchoubey, B., Lang, S., Winter, S., & Birbaumer, N. (2003). Cognitive processing in completely paralyzed patients with amyotrophic lateral sclerosis. *European Journal of Neurology, 10*, 551–558.

Kouider, S., de Gardelle, V., Sackur, J., & Dupoux, E. (2010). How rich is consciousness? The partial awareness hypothesis. *Trends in Cognitive Sciences, 14*, 301–307.

Kretschmann, H.-J., & Weinrich, W. (1992). *Cranial Neuroimaging and Clinical Neuroanatomy*. Stuttgart: Georg Thieme.

Kripke, S. A. (1980). *Naming and Necessity*. Cambridge, MA: Harvard University Press.

Lachhwani, D. P., & Dinner, D. S. (2003). Cortical stimulation in the definition of eloquent cortical areas. *Handbook of Clinical Neurophysiology, 3*, 273–286.

Lamme, V. A. F. (2003). Why visual attention and awareness are different. *Trends in Cognitive Sciences, 7*, 12–18.

Lamme, V. A. F. (2006). Towards a true neural stance on consciousness. *Trends in Cognitive Sciences, 10*, 494–501.

Lamme, V. A. F. (2010). How neuroscience will change our view on consciousness. *Cognitive Neuroscience, 1*, 204–220.

Lamme, V. A. F., & Roelfsema, P. R. (2000). The distinct modes of vision offered by feedforward and recurrent processing. *Trends in Neurosciences, 23*, 571–579.

Landsness, E., Bruno, A. A., Noirhomme, Q., Riedner, B., Gosseries, O., Schnakers, C., et al. (2010). Electrophysiological correlates of behavioural changes in vigilance in vegetative state and minimally conscious state. *Brain, 134*, 2222–2232.

Larkum, M. (2013). A cellular mechanism for cortical associations: An organizing principle for the cerebral cortex. *Trends in Neurosciences, 36*, 141–151.

Lazar, R. M., Marshall, R. S., Prell, G. D., & Pile-Spellman, J. (2010). The experience of Wernicke's aphasia. *Neurology, 55*, 1222–1224.

Leibniz, G. W. (1951). *Leibniz: Selections*. P. P. Wiener (Ed.). New York: Charles Scribner's Sons.

Lem, S. (1987). *Peace on Earth*. San Diego: Harcourt.

Lemaitre, A.-L., Herbet, G., Duffau, H., & Lafargue, G. (2018). Preserved metacognitive ability despite unilateral or bilateral anterior prefrontal resection. *Brain & Cognition, 120*, 48–57.

Lemon, R. N., & Edgley, S. A. (2010). Life without a cerebellum. *Brain, 133*, 652–654.

Levin, J. (2018). Functionalism. In E. N. Zalta (Ed.), *The Stanford Encyclopedia of Philosophy*. https://plato.stanford.edu/archives/fall2018/entries/functionalism/.

Levi-Strauss, C. (1969). *Raw and the Cooked: Introduction to a Science of Mythology*. New York: Harper & Row,

Lewis, D. (1983). Extrinsic properties. *Philosophical Studies, 44*, 197–200.

Li, F. F., VanRullen, R., Koch, C., & Perona, P. (2002). Rapid natural scene categorization in the near absence of attention. *Proceedings of the National Academy of Sciences, 99*, 9596–9601.

List, C. (2016). What is it like to be a group agent? *Noûs, 52*, 295–319.

Logan, G. D., & Crump, M. J. C. (2009). The left hand doesn't know what the right hand is doing. *Psychological Science, 20*, 1296–1300.

Loukola, O. J., Perry, C. J., Coscos, L., & Chittka, L. (2017). Bumblebees show cognitive flexibility by improving on an observed complex behavior. *Science, 355*, 833–836.

Luo, Q., Mitchell, D., Cheng, X., Mondillo, K., Mccaffrey, D., Holroyd, T., et al. (2009). Visual awareness, emotion, and gamma band synchronization. *Cerebral Cortex, 19*, 1896–1904.

Lutz, A., Jha, A. P., Dunne, J. D., & Saron, C. D. (2015). Investigating the phenomenological matrix of mindfulness-related practices from a neurocognitive perspective. *American Psychologist, 70*, 632–658.

Lutz, A., Slagter, H. A., Rawlings, N. B., Francis, A. D., Greischar, L. L., & Davidson, R. J. (2009). Mental training enhances attentional stability: Neural and behavioral evidence. *Journal of Neuroscience, 29*, 13418–13427.

Mack, A., & Rock, I. (1998). *Inattentional Blindness*. Cambridge, MA: MIT Press.

Macphail, E. M. (1998). *The Evolution of Consciousness*. Oxford: Oxford University Press.

Macphail, E. M. (2000). The search for a mental Rubicon. In C. Heyes and L. Huber (Eds.), *The Evolution of Cognition* (pp. 253–271). Cambridge, MA: MIT Press.

Markari, G. (2015). *Soul Machine—The Invention of the Modern Mind*. New York: W. W. Norton.

Markowitsch, H. J., & Kessler, J. (2000). Massive impairment in executive functions with partial preservation of other cognitive functions: The case of a young patient with severe degeneration of the prefrontal cortex. *Experimental Brain Research, 133*, 94–102.

Markram, H. (2006). The blue brain project. *Nature Reviews Neuroscience, 7*, 153–160.

Markram, H. (2012). The human brain project. *Scientific American, 306*, 50–55.

Markram, H., Muller, E., Ramaswamy, S., Reimann, M. W., Abdellah, M., Sanchez, C. A., et al. (2015). Reconstruction and simulation of neocortical microcircuitry. *Cell, 163*, 456–492.

Marks, L. (2017). What my stroke taught me. *Nautilus, 19*, 80–89.

Marr, D. (1982). *Vision*. San Francisco, CA: Freeman.

Mars, R. B., Sotiropoulos, S. N., Passingham, R. E., Sallet, J., Verhagen, L., Khrapitchev, A. A., et al. (2018). Whole brain comparative anatomy using connectivity blueprints. *eLife, 7*, e35237.

Marshall, W., Albantakis, L., & Tononi, G. (2016), Black-boxing and cause-effect power. *arXiv*, 1608.03461.

Marshall, W., Kim, H., Walker, S. I., Tononi, G., & Albantakis, L. (2017). How causal analysis can reveal autonomy in models of biological systems. *Philosophical Transactions of the Royal Society of London A, 375*. doi:10.1098/rsta.2016.0358.

Martin, J. T., Faulconer, A. Jr., & Bickford, R. G. (1959). Electroencephalography in anesthesiology. *Anesthesiology*, 20, 359–376.

Massimini, M., Ferrarelli, F., Huber, R., Esser, S. K., Singh, H., & Tononi, G. (2005). Breakdown of cortical effective connectivity during sleep. *Science, 309*, 2228–2232.

Massimini, M., & Tononi, G. (2018). *Sizing Up Consciousness*. Oxford: Oxford University Press.

Mataro, M., Jurado, M. A., García-Sanchez, C., Barraquer, L., Costa-Jussa, F. R., & Junque, C. (2001). Long-term effects of bilateral frontal brain lesion: 60 years after injury with an iron bar. *Archives of Neurology, 58*, 1139–1142.

Mayner, W. G. P., Marshall, W., Albantakis, L., Findlay, G., Marchman, R., & Tononi, G. (2018). PyPhi: A toolbox for integrated information theory. *PLOS Computational Biology, 14*(7), e1006343.

Melloni, L., Molina, C., Pena, M., Torres, D., Singer, W., & Rodriguez, E. (2007). Synchronization of neural activity across cortical areas correlates with conscious perception. *Journal of Neuroscience, 27*, 2858–2865.

Merker, B. (2007). Consciousness without a cerebral cortex: A challenge for neuroscience and medicine. *Behavioral and Brain Sciences, 30*, 63–81.

Metzinger, T. (2003). *Being No One: The Self Model Theory of Subjectivity*. Cambridge, MA: MIT Press.

Miller, G. A. (1956). The magical number seven, plus or minus two: some limits on our capacity for processing information. *Psychological Review, 63*, 81–97.

Miller, S., M. (Ed.). (2015). *The Constitution of Phenomenal Consciousness*. Amsterdam, Netherlands: Benjamins.

Milo, R., & Phillips, R. (2016). *Cell Biology by the Numbers*. New York: Garland Science.

Minsky, M. (1986). *The Society of Mind*. New York: Simon & Schuster.

Mitchell, R. W. (2009). Self awareness without inner speech: A commentary on Morin. *Consciousness & Cognition, 18*, 532–534.

Mitchell, T. J., Hacker, C. D., Breshears, J. D., Szrama, N. P., Sharma, M., Bundy, D. T., et al. (2013). A novel data-driven approach to preoperative mapping of functional cortex using resting-state functional magnetic resonance imaging. *Neurosurgery, 73*, 969–982.

Monti, M. M., Vanhaudenhuyse, A., Coleman, M. R., Boly, M., Pickard, J. D. Tshibanda, L., et al. (2010). Willful modulation of brain activity in disorders of consciousness. *New England Journal of Medicine, 362*, 579–589.

Mooneyham, B. W., & Schooler, J. W. (2013). The costs and benefits of mind-wandering: A review. *Canadian Journal of Experimental Psychology, 67*, 11–18.

Mørch, H. H. (2017). The integrated information theory of consciousness. *Philosophy Now, 121*, 12–16.

Mørch, H. H. (2018). Is the integrated information theory of consciousness compatible with Russellian panpsychism? *Erkenntnis*, 1–21. doi:10.1007/s10670-018-9995-6.

Morgan, C. L. (1894). *An Introduction to Comparative Psychology*. New York: Scribner.

Morin, A. (2009). Self-awareness deficits following loss of inner speech: Dr. Jill Bolte Taylor's case study. *Consciousness & Cognition, 18*, 524–529.

Mortensen, H. S., Pakkenberg, B., Dam, M., Dietz, R., Sonne, C., Mikkelsen, B., & Eriksen, N. (2014). Quantitative relationships in delphinid neocortex. *Frontiers in Neuroanatomy, 8*, 132.

Mudrik, L., Faivre, N., & Koch, C. (2014). Information integration without awareness. *Trends in Cognitive Sciences, 18*, 488–496.

Munk, M. H., Roelfsema, P. R., König, P., Engel, A. K., & Singer, W. (1996). Role of reticular activation in the modulation of intracortical synchronization. *Science, 272*, 271–274.

Murphy, M. J., Bruno, M. A., Riedner, B. A., Boveroux, P., Noirhomme, Q., Landsness, E. C., et al. (2011). Propofol anesthesia and sleep: A high-density EEG study. *Sleep, 34*, 283–291.

Nagel, T. (1974). What is it like to be a bat? *Philosophical Review, 83*, 435–450.

Nagel, T. (1979). *Mortal Questions*. Cambridge: Cambridge University Press.

Narahany, N. A., Greely, H. T., Hyman, H., Koch, C., Grady, C., Pasca, S. P., et al. (2018). The ethics of experimenting with human brain tissue. *Nature, 556*, 429–432.

Narikiyo, K., Mizuguchi, R., Ajima, A., Mitsui, S., Shiozaki, M., Hamanaka, H., et al. (2018). The claustrum coordinates cortical slow-wave activity. *bioRxiv*, doi:10.1101/286773.

Narr, K. L., Woods, R. P., Thompson, P. M., Szeszko, P., Robinson, D., Dimtcheva, T., & Bilder, R. M. (2006). Relationships between IQ and regional cortical gray matter thickness in healthy adults. *Cerebral Cortex, 17*, 2163–2171.

Newton, M. (2002). *Savage Girls and Wild Boys*. New York: Macmillan.

Nichelli, P. (2016). Consciousness and aphasia. In S. Laureys, O. Gosseries, & G. Tononi (Eds.), *The Neurology of Consciousness* (2nd ed., pp. 379–391). Amsterdam: Elsevier.

Niikawa, T. (2018). Moral status and consciousness. *Annals of the University of Bucharest–Philosophy, 67*, 235–257.

Nir, Y., & Tononi, G. (2010). Dreaming and the brain: From phenomenology to neurophysiology. *Trends in Cognitive Sciences, 14*, 88–100.

Nityananda, V. (2016). Attention-like processes in insects. *Proceedings of the Royal Society of London Series B: Biological Sciences, 283*, 20161986.

Norretranders, T. (1991). *The User Illusion: Cutting Consciousness Down to Size*. New York: Viking Penguin.

Noyes, R. Jr., & Kletti, R. (1976). Depersonalization in the face of life-threatening danger: A description. *Psychiatry, 39*, 19–27.

Odegaard, B., Knight, R. T., & Lau, H. (2017). Should a few null findings falsify prefrontal theories of conscious perception? *Journal of Neuroscience, 37*, 9593–9602.

Odier, D. (2005). *Yoga Spandakarika: The Sacred Texts at the Origins of Tantra*. Rochester, Vermont: Inner Traditions.

Oizumi, M., Albantakis, L., & Tononi, G. (2014). From the phenomenology to the mechanisms of consciousness: Integrated information theory 3.0. *PLOS Computational Biology, 10*, e1003588.

Oizumi, M., Tsuchiya, N., & Amari, S. I. (2016). Unified framework for information integration based on information geometry. *Proceedings of the National Academy of Sciences, 113*, 14817–14822.

O'Leary, M. A., Bloch, J. I., Flynn, J. J., Gaudin, T. J., Giallombardo, A., Giannini, N. P., et al. (2013). The placental mammal ancestor and the post–K-Pg radiation of placentals. *Science, 339*, 662–667.

O'Regan, J. K., Rensink, R. A., & Clark, J. J. (1999). Change-blindness as a result of "mudsplashes." *Nature, 398*, 34–35.

Owen, A. (2017). *Into the Gray Zone: A Neuroscientist Explores the Border between Life and Death*. London: Scribner.

Owen, M. (2018). Aristotelian causation and neural correlates of consciousness. *Topoi: An International Review of Philosophy*, 1–12. doi:10.1007/s11245-018-9606-9.

Palmer, S. (1999). *Vision Science: Photons to Phenomenology*. Cambridge, MA: MIT Press.

Parker, I. (2003). Reading minds. *New Yorker*, January 20.

Parvizi, J., & Damasio, A. (2001). Consciousness and the brainstem. *Cognition, 79*, 135–159.

Parvizi, J., Jacques, C., Foster, B. L., Withoft, N., Rangarajan, V., Weiner, K. S., & Grill-Spector, K. (2012). Electrical stimulation of human fusiform face-selective regions distorts face perception. *Journal of Neuroscience, 32*, 14915–14920.

Pasca, S. P. (2019). Assembling human brain organoids. *Science*, 363, 126–127.

Passingham, R. E. (2002). The frontal cortex: Does size matter? *Nature Neuroscience, 5*, 190–192.

Paul, L. K., Brown, W. S., Adolphs, R., Tyszka, J. M., Richards, L. J., Mukherjee, P., & Sherr, E. H. (2007). Agenesis of the corpus callosum: genetic, developmental and functional aspects of connectivity. *Nature Reviews Neuroscience, 8*, 287–299.

Pearl, J. (2000). *Causality: Models, Reasoning, and Inference*. Cambridge: Cambridge University Press.

Pearl, J. (2018). *The Book of Why: The New Science of Cause and Effect*. New York: Basic Books.

Pekala, R. J., & Kumar, V. K. (1986). The differential organization of the structure of consciousness during hypnosis and a baseline condition. *Journal of Mind and Behavior, 7*, 515–539.

Penfield, W. & Perot, P. (1963). The brain's record of auditory and visual experience: A final summary and discussion. *Brain, 86*, 595–696.

Penrose, R. (1989). *The Emperor's New Mind*. Oxford: Oxford University Press.

Penrose, R. (1994). *Shadows of the Mind*. Oxford: Oxford University Press.

Penrose, R. (2004). *The Road to Reality—A Complete Guide to the Laws of the Universe*. New York: Knopf.

Piersol, G. A. (1913). *Human Anatomy*. Philadelphia: J. B. Lippincott.

Pietrini, P., Salmon, E., & Nichelli, P. (2016). Consciousness and dementia: How the brain loses its self. In S. Laureys, O. Gosseries, &. G. Tononi (Eds.), *Neurology of Consciousness* (2nd ed., pp. 379–391). Amsterdam: Elsevier.

Pinto, Y., Haan, E. H. F., & Lamme, V. A. F. (2017). The split-brain phenomenon revisited: A single conscious agent with split perception. *Trends in Cognitive Sciences, 21,* 835–851.

Pinto, Y., Lamme, V. A. F., & de Haan, E. H. F. (2017). Cross-cueing cannot explain unified control in split-brain patients—Letter to the Editor. *Brain, 140,* 1–2.

Pinto, Y., Neville, D. A., Otten, M., Corballis, P. M., Lamme, V. A., de Haan, E. H., et al. (2017). Split brain: Divided perception but undivided consciousness. *Brain, 140,* 1231–1237.

Pitkow, X., & Meister, M. (2014). Neural computation in sensory systems. In M. S. Gazzaniga & G. R. Mangun (Eds.), *The Cognitive Neurosciences,* pp. 305–316. Cambridge, MA: MIT Press.

Pitts, M. A., Lutsyshyna, L. A., & Hillyard, S. A. (2018). The relationship between attention and consciousness: an expanded taxonomy and implications for "no-report" paradigms. *Philosophical Transactions of the Royal Society of London B, 373,* 20170348. doi:10.1098/rstb.2017.0348.

Pitts, M. A., Padwal, J., Fennelly, D., Martínez, A., & Hillyard, S. A. (2014). Gamma band activity and the P3 reflect post-perceptual processes, not visual awareness. *NeuroImage, 101,* 337–350.

Plomin, R. (2001). The genetics of *G* in human and mouse. *Nature Reviews Neuroscience, 2,* 136–141.

Pockett, S., & Holmes, M. D. (2009). Intracranial EEG power spectra and phase synchrony during consciousness and unconsciousness. *Consciousness and Cognition, 18,* 1049–1055.

Polak, M., & Marvan, T. (2018). Neural correlates of consciousness meet the theory of identity. *Frontiers in Psychology, 24.* doi:10.3389/fpsyg.2018.01269.

Popa, I., Donos, C., Barborica, A., Opris, I., Dragos, M., Mălîia, M., Ene, M., Ciurea, J., & Mîndruta, I. (2016). Intrusive thoughts elicited by direct electrical stimulation during stereo-electroencephalography. *Frontiers in Neurology, 7.* doi:10.3389/fneur.2016.00114.

Posner, J. B., Saper, C. B., Schiff, N. D., & Plum, F. (2007). *Plum and Posner's Diagnosis of Stupor and Coma.* New York: Oxford University Press.

Preuss, T. M. (2009). The cognitive neuroscience of human uniqueness. In M. S. Gazzaniga (Ed.), *The Cognitive Neuroscience* (pp. 49–64). Cambridge, MA: MIT Press.

Prinz, J. (2003). A neurofunctional theory of consciousness. In A. Brook & K. Akins (Eds.), *Philosophy and Neuroscience.* Cambridge: Cambridge University Press.

Puccetti, R. (1973). *The Trial of John and Henry Norton.* London: Hutchinson.

Quadrato, G., Nguyen, T., Macosko, E. Z., Sherwood, J. L., Yang, S. M., Berger, D. R., et al. (2017). Cell diversity and network dynamics in photosensitive human brain organoids. *Nature, 545*, 48–53.

Railo, H., Koivisto, M., & Revonsuo, A. (2011). Tracking the processes behind conscious perception: A review of event-related potential correlates of visual consciousness. *Consciousness and Cognition, 20*, 972–983.

Ramsoy, T. Z., & Overgaard, M. (2004). Introspection and subliminal perception. *Phenomenology and the Cognitive Sciences, 3*, 1–23.

Rangarajan, V., Hermes, D., Foster, B. L., Weinfer, K. S., Jacques, C., Grill-Spector, K., & Parvizi, J. (2014). Electrical stimulation of the left and right human fusiform gyrus causes different effects in conscious face perception. *Journal of Neuroscience, 34*, 12828–12836.

Rangarajan, V., & Parvizi, J. (2016). Functional asymmetry between the left and right human fusiform gyrus explored through electrical brain stimulation. *Neuropsychologia, 83*, 29–36.

Rathi, Y., Pasternak, O., Savadjiev, P., Michailovich, O., Bouix, S., Kubicki, M., et al. (2014) Gray matter alterations in early aging: A diffusion magnetic resonance imaging study. *Human Brain Mapping, 35*, 3841–3856.

Rauschecker, A. M., Dastjerdi, M., Weiner, K. S., Witthoft, N., Chen, J., Selimbeyoglu, A., & Parvizi, J. (2013). Illusions of visual motion elicited by electrical stimulation of human MT complex. *PLoS ONE, 6*. doi:10.1371/journal.pone.0021798.

Ray, S., & Maunsell, J. H. (2011). Network rhythms influence the relationship between spike-triggered local field potential and functional connectivity. *Journal of Neuroscience, 31*, 12674–12682.

Reardon, S. (2017). A giant neuron found wrapped around entire mouse brain. *Nature, 543*, 14–15.

Reimann, M. W., Anastassiou, C. A., Perin, R., Hill, S., Markram, H., & Koch, C. (2013). A biophysically detailed model of neocortical local field potentials predicts the critical role of active membrane currents. *Neuron, 79*, 375–390.

Rensink, R.A., O'Regan, J.K., & Clark, J.J. (1997). To see or not to see: The need for attention to perceive changes in scenes. *Psychological Sciences, 8*, 368–373.

Rey, G. (1983). A Reason for doubting the existence of consciousness. In R. Davidson, G. Schwarz, and D. Shapiro (Eds.), *Consciousness and Self-Regulation: Advances in Research and Theory* (Vol. 3). New York: Plenum Press.

Rey, G. (1991). Reasons for doubting the existence of even epiphenomenal consciousness. *Behavioral & Brain Science, 14*, 691–692.

Ricard, M., Lutz, A., & Davidson, R. J. (2014). Mind of the meditator. *Scientific American, 311*, 38–45.

Ridley, M. (2006). *Francis Crick*. New York: HarperCollins.

Rodriguez, E., George, N., Lachaux, J.-P., Martinerie, J., Renault, B., & Varela, F.J. (1999). Perception's shadow: Long-distance synchronization of human brain activity. *Nature, 397*, 430–433.

Roelfsema, P. R., Engel, A. K., König, P., & Singer, W. (1997). Visuomotor integration is associated with zero time-lag synchronization among cortical areas. *Nature, 385*, 157–161.

Romer, A. S., & Sturges, T. S. (1986). *The Vertebrate Body* (6th ed.). Philadelphia: Saunders College.

Roth, G., & Dicke, U. (2005). Evolution of the brain and intelligence. *Trends in Cognitive Sciences, 9*, 250–257.

Rowe, T. B., Macrini, T. E., & Luo, Z.-X. (2011). Fossil evidence on origin of the mammalian brain. *Science, 332*, 955–957.

Rowlands, M. (2009). *The Philosopher and the Wolf*. New York: Pegasus Books.

Rowlatt, P. (2018). *Mind, a Property of Matter*. London: Ionides Publishing.

Ruan, S., Wobbrock, J. O., Liou, K., Ng, A., & Landay, J. A. (2017). Comparing speech and keyboard text entry for short messages in two languages on touchscreen phones. In *Proceedings of ACM Interactive, Mobile, Wearable and Ubiquitous Technologies*. doi:10.1145/3161187.

Ruff, C. B., Trinkaus, E., & Holliday, T. W. (1997). Body mass and encephalization in Pleistocene Homo. *Nature, 387*, 173–176.

Russell, B. (1927). *The Analysis of Matter*. London: George Allen & Unwin.

Russell, R., Duchaine, B., & Nakayama, K. (2009). Super-recognizers: People with extraordinary face recognition ability. *Psychonomic Bulletin and Review, 16*, 252–257.

Rymer, R. (1994). *Genie: A Scientific Tragedy* (2nd ed.). New York: Harper Perennial.

Sacks, O. (2010). *The Mind's Eye*. New York: Knopf.

Sacks, O. (2017). *The River of Consciousness*. New York: Knopf.

Sadava, D., Hillis, D. M., Heller, H. C., & Berenbaum, M. R. (2011). *Life: The Science of Biology* (9th ed.). Sunderland, MA: Sinauer and W.H. Freeman.

Sandberg, K., Timmermans, B., Overgaard, M., & Cleeremans, A. (2010). Measuring consciousness: Is one measure better than the other? *Consciousness and Cognition, 19*, 1069–1078.

Sanes, J. R., & Masland, R. H. (2015). The types of retinal ganglion cells: Current status and implications for neuronal classification. *Annual Review of Neuroscience, 38*, 221–246.

Saper, C. B., & Fuller, P. M. (2017). Wake-sleep circuitry: An overview. *Current Opinion in Neurobiology, 44*, 186–192.

Saper, C. B., Scammell, T. E., & Lu, J. (2005). Hypothalamic regulation of sleep and circadian rhythms. *Nature, 437*, 1257–1263.

Saposnik, G., Bueri, J. A., Mauriño, J., Saizar, R., & Garretto, N. S. (2000). Spontaneous and reflex movements in brain death. *Neurology, 54*, 221.

Sasai, S., Boly, M., Mensen, A., & Tononi, G. (2016). Functional split brain in a driving/listening paradigm. *Proceedings of the National Academy of Sciences, 113*, 14444–14449.

Scammell, T. E., Arrigoni, E., & Lipton, J. O. (2016). Neural circuitry of wakefulness and sleep. *Neuron, 93*, 747–765.

Schalk, G., Kapeller, C., Guger, C., Ogawa, H., Hiroshima, S., Lafer-Sousa, R., et al. (2017). Facephenes and rainbows: Causal evidence for functional and anatomical specificity of face and color processing in the human brain. *Proceedings of the National Academy of Sciences, 114*, 12285–12290.

Schartner, M. M., Carhart-Harris, R. L., Barrett, A. B., Seth, A. K., & Muthukumaraswamy, S. D. (2017). Increased spontaneous MEG signal diversity for psychoactive doses of ketamine, LSD and psilocybin. *Scientific Reports, 7*, 46421.

Schiff, N. D. (2013). Making waves in consciousness research. *Science Translational Medicine, 5*, 1–3.

Schiff, N. D., & Fins, J. J. (2016). Brain death and disorders of consciousness. *Current Biology, 26*, R572–R576.

Schimmack, U., Heene, M., & Kesavan, K. (2017). Reconstruction of a train wreck: How priming research went off the rails. https://replicationindex.wordpress.com/2017/02/02/reconstruction-of-a-train-wreck-how-priming-research-went-of-the-rails/.

Schmidt, E. M., Bak, M. J., Hambrecht, F. T., Kufta, C. V., O'Rourke, D. K., & Vallabhanath, P. (1996). Feasibility of a visual prosthesis for the blind based on intracortical microstimulation of the visual cortex. *Brain, 119*, 507–522.

Schooler, J. W., & Melcher, J. (1995). The ineffability of insight. In S. M. Smith, T. B. Ward, & R. A. Finke (Eds.), *The Creative Cognition Approach* (pp. 97–134). Cambridge, MA: MIT Press.

Schooler, J. W., Ohlsson, S., & Brooks, K. Thoughts beyond words: When language overshadows insight. *Journal of Experimental Psychology—General, 122*, 166–183.

Schopenhauer, A. (1813). *On the Fourfold Root of the Principle of Sufficient Reason.* (Hillebrand, K., Trans.; rev. ed., 1907). London: George Bell & Sons.

Schrödinger, E. (1958). *Mind and Matter.* Cambridge: Cambridge University Press.

Schubert, R., Haufe, S., Blankenburg, F., Villringer, A., & Curio, G. (2008). Now you'll feel it—now you won't: EEG rhythms predict the effectiveness of perceptual masking. *Journal of Cognitive Neuroscience, 21,* 2407–2419.

Searle, J. R. (1980). Minds, brains, and programs. *Behavioral and Brain Sciences, 3,* 417–424.

Searle, J. R. (1992). *The Rediscovery of the Mind.* Cambridge, MA: MIT Press.

Searle, J. R. (1997). *The Mystery of Consciousness.* New York: New York Review Books.

Searle, J. R. (2013a). Can information theory explain consciousness? *New York Review of Books,* January 10, 54–58.

Searle, J. R. (2013b). Reply to Koch and Tononi. *New York Review of Books,* March 7.

Seeley, T. D. (2000). *Honeybee Democracy.* Princeton: Princeton University Press.

Selimbeyoglu, A., & Parvizi, J. (2010). Electrical stimulation of the human brain: Perceptual and behavioral phenomena reported in the old and new literature. *Frontiers in Human Neuroscience, 4.* doi:10.3389/fnhum.2010.00046.

Sender, R., Fuchs, S., & Milo, R. (2016). Revised estimates for the number of human and bacteria cells in the body. *PLOS Biology, 14,* e1002533.

Sergent, C., & Dehaene, S. (2004). Is consciousness a gradual phenomenon? Evidence for an all-or-none bifurcation during the attentional blink. *Psychological Science, 15,* 720–728.

Seth, A. K. (2015). Inference to the best prediction: A reply to Wanja Wiese. In T. Metzinger & J. M. Windt (Eds.), *OpenMIND* (p. 35). Cambridge, MA: MIT Press.

Seung, S. (2012). *Connectome: How the Brain's Wiring Makes Us Who We Are.* New York: Houghton Mifflin Harcourt.

Shanahan, M. (2015). Ascribing consciousness to artificial intelligence. *arXiv,* 1504.05696v2.

Shanks, D. R., Newell, B. R., Lee, E. H., Balakrishnan, D., Ekelund, L., Cenac, Z., Kavvadia, F., & Moore, C. (2013). Priming intelligent behavior: An elusive phenomenon. *PLOS One, 8*(4), e56515.

Shear, J. (Ed.). (1997). *Explaining Consciousness: The Hard Problem.* Cambridge, MA: MIT Press.

Shewmon, D. A. (1997). Recovery from "brain death": A neurologist's apologia. *Linacre Quarterly, 64,* 30–96.

Shipman, P. (2015). *The Invaders: How Humans and Their Dogs Drove Neanderthals to Extinction.* Cambridge, MA: Harvard University Press.

Shubin, N. (2008). *Your Inner Fish—A Journey into the 3.5-billion Year History of the Human Body.* New York: Vintage.

Siclari, F., Baird, B., Perogamvros, L., Bernadri, G., LaRocque, J. J., Riedner, B., Boly, M., Postle, B. R., & Tononi, G. (2017). The neural correlates of dreaming. *Nature Neuroscience, 20,* 872–878.

Simons, D. J., & Chabris, C.F. (1999). Gorillas in our midst: Sustained inattentional blindness for dynamic events. *Perception, 28,* 1059–1074.

Simons, D. J., & Levin, D. T. (1997). Change blindness. *Trends in Cognitive Sciences, 1,* 261–267.

Simons, D. J., & Levin, D. T. (1998) Failure to detect changes to people during a real-world interaction. *Psychonomic Bulletin & Review, 5,* 644–649.

Singer, P. (1975). *Animal Liberation.* New York: HarperCollins.

Singer, P. (1994). *Rethinking Life and Death.* New York: St. Martin's Press.

Skrbina, D. F. (2017). *Panpsychism in the West* (Rev. ed.). Cambridge, MA: MIT Press.

Sloan, S. A., Darmanis, S., Huber, N., Khan, T. A., Birey, F., Caneda, C., & Paşca, S. P. (2017). Human astrocyte maturation captured in 3D cerebral cortical spheroids derived from pluripotent stem cells. *Neuron, 95,* 779–790.

Snaprud, P. (2018). The consciousness wager. *New Scientist,* June 23, 26–29.

Sperry, R. W. (1974). Lateral specialization in the surgically separated hemispheres. In F. O. Schmitt and F.G. Worden (Eds.), *Neuroscience 3rd Study Program.* Cambridge, MA: MIT Press.

Stickgold, R., Malaia, A., Fosse, R., Propper, R., & Hobson, J.A. (2001). Brain-mind states: Longitudinal field study of sleep/wake factors influencing mentation report length. *Sleep, 24,* 171–179.

Strawson, G. (1994). *Mental Reality.* Cambridge, MA: MIT Press.

Strawson, G. (2018). The consciousness deniers. *New York Review of Books,* March 13.

Sullivan, P. R. (1995). Content-less consciousness and information-processing theories of mind. *Philosophy, Psychiatry, and Psychology, 2,* 51–59.

Super, H., Spekreijse, H., & Lamme, V. A. F. (2001). Two distinct modes of sensory processing observed in monkey primary visual cortex. *Nature Neuroscience, 4,* 304–310.

Takahashi, N., Oertner T. G., Hegemann, P., & Larkum, M.E. (2016). Active cortical dendrites modulate perception. *Science, 354,* 1587–1590.

Taneja, B., Srivastava, V., & Saxena, K. N. (2012). Physiological and anaesthetic considerations for the preterm neonate undergoing surgery. *Journal of Neonatal Surgery, 1,* 14.

Tasic, B., Yao, Z., Graybuck, L., T., Smith, K. A., Nguyen, T. N. Bertagnolli, D., Giddy, J. et al. (2018). Shared and distinct transcriptomic cell types across neocortical areas. *Nature, 563,* 72–78.

Teilhard de Chardin, P. (1959). *The Phenomenon of Man.* New York: Harper.

Tegmark, M. (2000). The importance of quantum decoherence in brain processes. *Physical Review E, 61,* 4194–4206.

Tegmark, M. (2014). *Our Mathematical Universe: My Quest for the Ultimate Nature of Reality.* New York: Alfred Knopf.

Tegmark, M. (2015). Consciousness as a state of matter. *Chaos, Solitons & Fractals, 76,* 238–270.

Tegmark, M. (2016). Improved measures of integrated information. *PLOS Computational Biology, 12*(11), e1005123.

Teresi, D. (2012). *The Undead—Organ Harvesting, the Ice-Water Test, Beating-Heart Cadavers—How Medicine Is Blurring the Line between Life and Death.* New York: Pantheon Books.

Thompson, E. (2007). *Mind in Life: Biology, Phenomenology, and the Sciences of the Mind.* Cambridge, MA: Harvard University Press.

Tononi, G. (2012). Integrated information theory of consciousness: An updated account. *Archives Italiennes de Biology, 150,* 290–326.

Tononi, G., Boly, M., Gosseries, O., & Laureys, S. (2016). The neurology of consciousness. In S. Laureys, O. Gosseries, & G. Tononi (Eds.), *The Neurology of Consciousness* (2nd ed., pp. 407–461). Amsterdam: Elsevier.

Tononi, G., Boly, M., Massimini, M., & Koch, C. (2016). Integrated information theory: From consciousness to its physical substrate. *Nature Reviews Neuroscience, 17,* 450–461.

Tononi, G., & Koch, C. (2015). Consciousness: Here, there and everywhere? *Philosophical Transactions of the Royal Society of London B, 370,* 20140167.

Travis, S. L., Dux, P. E., & Mattingley, J. B. (2017). Re-examining the influence of attention and consciousness on visual afterimage duration. *Journal of Experimental Psychology: Human Perception and Performance, 43,* 1944–1949.

Treisman, A. (1996). The binding problem. *Current Opinions in Neurobiology, 6,* 171–178.

Trujillo, C. A., Gao, R., Negraes, P. D., Chaim, I. A., Momissy, A., Vandenberghe, M., Devor, A., Yeo, G. W., Voytek, B., & Muotri, A. R. (2018). Nested oscillatory dynamics in cortical organoids model early human brain network development. *bioRxiv,* doi:10.1101/358622.

Truog, R. D., & Miller, F. G. (2014). Changing the conversation about brain death. *American Journal of Bioethics, 14,* 9–14.

Tsuchiya, N., & Koch, C. (2005). Continuous flash suppression reduces negative afterimages. *Nature Neuroscience, 8,* 1096–101.

Tsuchiya, N., Taguchi, S., & Saigo, H. (2016). Using category theory to assess the relationship between consciousness and integrated information theory. *Neuroscience Research, 107,* 1–7.

Turing, A. (1950). Computing machinery and intelligence. *Mind, 59,* 433–460.

Tyszka, J. M., Kennedy, D. P., Adolphs, R., & Paul, L. K. (2011). Intact bilateral resting-state networks in the absence of the corpus callosum. *Journal of Neuroscience, 31,* 15154–15162.

VanRullen, R. (2016). Perceptual cycles. *Trends in Cognitive Sciences, 20,* 723–735.

VanRullen, R., & Koch, C. (2003). Is perception discrete or continuous? *Trends in Cognitive Sciences, 7,* 207–213.

VanRullen, R., Reddy, L., & Koch, C. (2010). A motion illusion reveals the temporally discrete nature of visual awareness. In R. Nijhawam & B. Khurana (Eds.), *Space and Time in Perception and Action* (pp. 521–535). Cambridge: Cambridge University Press.

van Vugt, B., Dagnino, B., Vartak, D., Safaai, H., Panzeri, S., Dehaene, S., & Roelfsema, P. R. (2018). The threshold for conscious report: Signal loss and response bias in visual and frontal cortex. *Science, 360,* 537–542.

Varki, A., & Altheide, T. K. (2005). Comparing the human and chimpanzee genomes: Searching for needles in a haystack. *Genome Research, 15,* 1746–1758.

Vilensky, J. A. (Ed.). (2011). *Encephalitis Lethargica—During and After the Epidemic.* Oxford: Oxford University Press.

Volz, L. J., & Gazzaniga, M. S. (2017). Interaction in isolation: 50 years of insights from split-brain research. *Brain, 140,* 2051–2060.

Volz, L. J., Hillyard, S. A., Miler, M. B., & Gazzaniga, M. S. (2018). Unifying control over the body: Consciousness and cross-cueing in split-brain patients. *Brain, 141,* 1–3.

von Arx, S. W., Müri, R. M., Heinemann, D., Hess, C. W., & Nyffeler, T. (2016). Anosognosia for cerebral achromatopsia: A longitudinal case study. *Neuropsychologia, 48*, 970–977.

von Bartheld, C. S., Bahney, J., & Herculano-Houzel, S. (2016). The search for true numbers of neurons and glial cells in the human brain: A review of 150 years of cell counting. *Journal of Comparative Neurology, 524*, 3865–3895.

Vyazovskiy, V. V., Olcese, U., Hanlon, E. C., Nir, Y., Cirelli, C., & Tononi, G. (2011). Local sleep in awake rats. *Nature, 472*, 443–447.

Wallace, D.F. (2004). Consider the lobster. *Gourmet* (August): 50–64.

Walloe, S., Pakkenberg, B., & Fabricius, K. (2014). Stereological estimation of total cell numbers in the human cerebral and cerebellar cortex. *Frontiers in Human Neuroscience, 8*, 508–518.

Wan, X., Nakatani, H., Ueno, K., Asamizuya, T., Cheng, K., & Tanaka, K. (2011). The neural basis of intuitive best next-move generation in board game experts. *Science, 331*, 341–346.

Wan, X., Takano, D., Asamizuya, T., Suzuki, C., Ueno, K., Cheng, K., et al. (2012). Developing intuition: Neural correlates of cognitive-skill learning in caudate nucleus. *Journal of Neuroscience, 32*, 492–501.

Wang, Q., Ng, L., Harris, J. A., Feng, D., Li, Y., Royall, J. J., et al. (2017a). Organization of the connections between claustrum and cortex in the mouse. *Journal of Comparative Neurology, 525*, 1317–1346.

Wang, Y., Li, Y., Kuang X. Rossi, B., Daigle, T. L., Madisen, L., Gu, H., Mills, M., Gray, L., Tasic, B., Zhou, Z., et al. (2017b). Whole-brain reconstruction and classification of spiny claustrum neurons and L6b-PCs of Gnb4 Tg mice. Poster presentation at Society of Neuroscience, 259.02. Washington, DC.

Ward, A. F. & Wegner, D. M. (2013). Mind-blanking: When the mind goes away. *Frontiers in Psychology, 27*. doi:10.3389/fpsyg.2013.00650.

Whitehead, A. (1929). *Process and Reality*. New York: Macmillan.

Wigan, A. L. (1844). Duality of the mind, proved by the structure, functions, and diseases of the brain. *Lancet, 1*, 39–41.

Wigner, E. (1967). *Symmetries and Reflections: Scientific Essays*. Bloomington: Indiana University Press.

Williams, B. (1978). *Descartes: The Project of Pure Enquiry*. New York: Penguin.

Wimmer, R. A., Leopoldi, A., Aichinger, M., Wick, N., Hantusch, B., Novatchkova, M., Taubenschmid, J., Hämmerle, M., Esk, C., Bagley, J. A., Lindenhofer, D., et al.

(2019). Human blood vessel organoids as a model of diabetic vasculopathy. *Nature, 565,* 505–510.

Winawer, J., & Parvizi, J. (2016). Linking electrical stimulation of human primary visual cortex, size of affected cortical area, neuronal responses, and subjective experience. *Neuron, 92,* 1–7.

Winslade, W. (1998). *Confronting Traumatic Brain Injury.* New Haven: Yale University Press.

Wohlleben, P. (2016). *The Hidden Life of Trees.* Vancouver: Greystone.

Woolhouse, R. S., & Francks, R. (Eds.). (1997). *Leibniz's "New System" and Associated Contemporary Texts.* Oxford: Oxford University Press.

Wyart, V., & Tallon-Baudry, C. (2008). Neural dissociation between visual awareness and spatial attention. *Journal of Neuroscience, 28,* 2667–2679.

Yu, F., Jiang, Q. J., Sun, X. Y., & Zhang, R. W. (2014). A new case of complete primary cerebellar agenesis: Clinical and imaging findings in a living patient. *Brain, 138,* 1–5.

Zadra, A., Desautels, A., Petit, D., & Montplaisir, J. (2013). Somnambulism: Clinical aspects and pathophysiological hypotheses. *Lancet Neurology, 12,* 285–294.

Zanardi, P., Tomka, M., & Venuti, L. C. (2018). Quantum integrated information theory. *arXiv,* 1806.01421v1.

Zeki, S. (1993). *A Vision of the Brain.* Oxford: Oxford University Press.

Zeng, H., & Sanes, J. R. (2017). Neuronal cell-type classification: challenges, opportunities and the path forward. *Nature Reviews Neuroscience, 18,* 530–546.

Zimmer, C. (2004). *Soul Made Flesh: The Discovery of the Brain.* New York: Free Press.

Zurek, W. H. (2002). Decoherence and the transition from quantum to classical-revisited. *Los Alamos Science, 27,* 86–109.

Index

Note: an "f" following a page number indicates a figure.

Aaronson, Scott, 152–153, 201n15
Abductive reasoning, 12–13, 26, 155, 158, 178n3
Access consciousness, 19, 179n11
Achromatopsia, 60–61, 181n9
Action potential, 42, 98, 117, 138, 203n11
Adversarial collaboration, 204n16
Agnosia, 60, 187n8
Akinetopsia, 60, 187n8
Alethiometer, 131
Alien hand syndrome, 197n6
AlphaGo Zero, 132, 151
Alpha waves, 44–45
Amobarbital, 196n2
Anencephaly, 93, 171
Anesthesia
  BIS monitor and, 99, 195n15
  cerebral organoids and, 128
  electroencephalography (EEG) and, 99, 101–102, 194n12
  interoperative recall and, 93
  measurement and, 93–94, 99, 101–102, 104
  pain and, 47, 62, 93–94
  propofol and, 101–102
Animal consciousness, 25–31
  behavioral continuity and, 26–28
  cognition and, 26, 31, 170–171, 207n4
  color and, 181n9
  empathy and, 172–173
  experience and, 28–31, 155–158
  intelligence and, 125–127, 201n12
  neocortex and, 27, 157–158, 180n6
  nervous system size and, 201n12
  Nonhuman Rights Project and, 173
  pain and, 26–29, 209n4
  sentience and, 25, 29, 157, 159, 170–172
  tree of life and, 155–158
  vivisection and, 26, 180n2
Anosognosia, 60, 187nn8,9
Aphasia, 29–30, 61
Aristotle, xi, 39, 85, 119, 138, 175n3, 183n3, 185n14, 192n9
Artificial intelligence (AI), 13, 132–134, 139, 141, 167, 191n1
Attention, 36–38, 99
  consciousness and, 36–38, 183n8
  experience and, 10, 13, 21, 33, 63, 204n16
  forms of, 36–37
  gist and, 38
  hemodynamic activity and, 188n19
  inattentional blindness and, 37, 183n7, 204n15
  integrated information theory (IIT) and, 204n16
  selective, 37–38, 99, 176n10, 183n9, 195n14, 207n4
  visual cortex and, 194n13
Awareness, 1. *See also* Consciousness

Axioms of integrated information theory (IIT), 75–77
consciousness exists intrinsically, 1–4, 74–75, 80–82
consciousness is definite, 9, 75, 86–87
consciousness is one (integrated), 8–9, 75, 85–86
consciousness is the specific way it is (informative), 8, 75, 84–85
consciousness is structured, 6–8, 75, 82–84
Ayahuasca, 46

Background conditions for consciousness, 51, 54–55, 82, 89, 107
Basal ganglia, 58, 63–64, 112, 186n4
Being, 5, 36, 115, 130, 142–144, 172
Berger, Hans, 44
Bernese mountain dog, 5–6, 164
Binding problem, 193n13, 194n13
Birbaumer, Niels, 194n11
Bispectral index (BIS) monitors, 99, 195n15
Blackboard architecture, 139
Blackmore, Susan, 177n12
Blade Runner (film), 129
Blind Café, 8
Blinks, 20, 95, 97
Blood flow, 49, 55, 95
Blue Brain Project, 203n12
Bolte Taylor, Jill, 29–30
Brain. *See also specific area*
coma and, 23, 47, 55, 59, 96
connectome and, 122, 138, 151, 206n17
conscious states and, 43–47
death and, 39, 43, 94, 96
encephalization quotient and, 181n7
group mind and, 111, 163–167
historical perspective on, 39–43
IQ and, 105, 125, 151, 188n15, 200n9, 201n12
multiple minds and, 112–113
organizational levels of, 196n21
physicalism and, 48–49
quantum mechanics and, 67–69
seizures and, 30, 40, 61, 63, 65, 105, 187n11, 189n23, 195n14
split, 30, 105–111, 163, 165, 182n15, 197nn3,5
Brain-as-computer metaphor, 130, 134–137
Brain-bridging technology, 108–111, 162, 165
Brainstem
consciousness and, 53–55
damage to, 54
death and, 54
reflex and, 20
spinal cord and, 53, 58
Brain waves
alpha, 44
electroencephalography (EEG) and, 44–46, 179n17
gamma, 44, 98–99, 194n13, 195n14, 199n23
sleep and, 23, 46
Broca's area, 62f, 66, 108, 112–113, 188n14, 196n2
Brodmann areas, 187n12, 188nn14,15,18
Buddhism, 7, 25, 76, 114, 173

Calculus ratiocinator, 130–131
Causal powers, 81–82
critical role of, 191n1
experience and, 10–11, 141–154, 158–160, 166–167, 178n1, 187n10, 205n12, 206n15, 209n3
extrinsic, 149, 192n9, 208n18
granularity of, 104, 196n21
integrated information theory (IIT) and, 74, 79–82, 88, 104, 116, 140–142, 145, 147, 153, 159–160, 166–167, 191n3, 196n21, 205n12

intrinsic, 79–89, 104, 116–117, 122, 134, 141, 145–150, 158, 191nn1,3, 192.n9
physical substrate and, 192n12
Whole and, 79–89, 191nn1,3,12, 192n9
Cause-effect structure. *See* Maximally irreducible cause-effect structure
Cell types, 42, 136–137, 184n6
Central identity of IIT, 80f, 87–89, 106
Cerebellar cognitive affective syndrome, 186n5
Cerebellum
experience and, 55–58, 63, 67, 158, 162, 186nn4,6, 207n3
feedback and, 55, 58
lack of, 55–58, 186n6
Purkinje cells and, 56–58
Cerebral organoids, 126–128, 201n14
consciousness and, 127–128, 152, 154, 170
Cerf, Moran, 197n8
Chalmers, David, 48, 73–74, 94, 185n14, 190n2, 193n4, 195n19, 208n13
Change blindness, 183n7, 204n15
Changeux, Jean-Pierre, 139
Chardin, Teilhard de, 161
Chess, 120, 132, 151, 202n10
Childhood amnesia, 181n11
Childhood's End? (Clarke), 111
Church, Alonzo, 131
Churchland, Patricia and Paul, 3
City of God (Saint Augustine), 2
Claustrum, 58, 66–67, 112, 189n25, 198n9
Cloud of Unknowing, The (Anonymous), 114
Cognitive ability, 125, 171
Color
achromatopsia and, 60–61, 181n9
animal consciousness and, 181n9
brain and, 43, 46, 48–49
experience and, 5–8, 14, 19, 43, 46, 48–49, 60–61, 64, 187nn8,9
Lilac chaser and, 16–17
Coma
brain and, 23, 47, 55, 59
electroencephalography (EEG) and, 99, 101
vegetative state (VS) and, 96, 103
Combination problem, 163, 192n7, 208n17
Computationalism
algorithms and, 5, 79, 101, 130, 132, 140, 149, 151, 193n14
artificial intelligence (AI) and, 13, 132–134, 139, 141, 167, 191n1
blackboard architecture and, 139
brain-as-computer metaphor and, 130, 134–137
Chinese room argument and, 205n12
cognition and, 129, 133–134, 139
consciousness, 140, 204n17
as faith of information age, 130–132
global workspace theory and, 139–140
information processing and, 132, 134
intelligence and, 129, 132–134, 139, 176n9, 209n5
machine learning and, 103, 129, 132, 135, 144, 184n10
mind-as-software mythos and, 129–130, 134f
Searle and, 205n12
Turing and, 131–134, 137t, 147, 150–151, 202n3, 206n16
Computers
brain as, 130, 134–137
emulation and, 138, 203n13
expander graphs and, 152–154, 206n19
fake empathy and, 209n5
functionalism and, 128, 143–150
futility of mind uploading and, 148–151

Computers (cont.)
integrated information theory (IIT) and, 74, 101, 141–148, 151–154, 166, 171
as lacking experience, 38, 141–154
logic gates and, 82–87, 142–150, 153, 206n20
minuscule intrinsic existence and, 144–148
neuromorphic electronic hardware and, 150
reducible, 141–148, 153–154
Turing machine and, 147, 150–151, 206n16
von Neumann and, 133, 145, 150, 171, 190n28
Confidence, subjective, 17–18, 22, 63, 93, 179n8, 188n19
Connectome, 122, 138, 151, 206n17
Consciousness
across lifespan, 162
attention and, 36–38, 183n8
brainstem and, 53–55
criteria for, 1–10
definition of, 1–2, 5–6
denial of, 3–4, 161
depths of, 17–19
disorders of, 23, 94–95, 193n6
epiphenomenalism, 120–121
experience and, 5–6 (*see also* Experience)
first-person perspective and, 11, 17, 22, 182n13
global workspace theory and, 139–140
group mind and, 111, 163–167
importance of, 169–173
intelligence-consciousness (I-C) plane, 124–128, 154
language and, 28–31, 105, 181nn11,12
limits of behavioral methods and, 22–24
loss of cerebellum and, 55–58
others' minds and, 13–17
phenomenal, 1, 6, 8, 19, 73, 120, 126, 147, 155, 165–166, 171, 177n15, 179n11
physical substrate of, 87
quantum mechanics and, 67–69, 73, 77, 151–152, 161, 190nn28–30, 208n11
self-, 17, 126, 182n13
spandrel, 121
states of, 43–47
subjective, 1, 3–5, 11, 14, 17–18, 22, 42–43, 57, 59, 76, 116, 119, 130, 133, 155, 177nn17,18, 179n8, 190n28
third-person properties and, 11, 14, 17, 22
tree of life and, 31, 155–158, 169, 173, 206n2
ubiquitousness of, 155–167
vegetative state and, 47, 96, 102–103, 171, 193n6
Continuous flash suppression, 183n8
Conversion disorder, 113
Corpus callosum, 30, 105, 107f, 108, 110f, 196n1, 197n3
Cortex
anencephaly and, 93, 171
consciousness residing in, 58–59
electrical brain stimulation (EBS) and, 64–67
eloquent vs. non-eloquent, 61–62
inactive vs. inactivated, 116–117
neural correlates of consciousness (NCC) and, 61, 63–64, 67, 188n20
number of neurons in, 180n5
pure consciousness and, 114–117
silent, 114–116
Cortical carpets, 128, 152–154, 201n15
Creativity, 34–36, 64, 171, 198n13
Crick, Francis
claustrum and, 66–67, 189n25
death of, 185n13
executive summary hypothesis and, 124

Freud and, 175n5
gamma-range synchronization and, 98, 194n13
neural correlates of consciousness (NCC) and, 47–49, 185n14
unconscious homunculus and, 35f, 64
zombie agents and, 20, 120
Crown of thorns neurons, 198n9

Dalai Lama, 25, 75–76
Darwin, Charles, 12, 123, 157, 160
Death, 156
beating heart cadaver, 194n9
brain and, 39, 43, 54, 94, 96
clinical rules for, 194n9
coma and, 96
functionalism and, 119, 121, 123
integrated information theory (IIT) and, 151
"life" after, 96
near-death experiences and, 115
Schiavo and, 23
Deduction, 2, 190n4
Deep Blue, 151
DeepMind, 132, 151, 152f
Dehaene, Stanislas, 139, 178n5, 179n8, 183n9, 187n10, 188n19, 195n16, 204n15
Dementia, 59, 103–104, 171, 193n3
Dendrites, 42, 137, 190n27
Dennett, Daniel, 3–4, 175n5
Denton, D., 176n11
Descartes, René, 114
animal consciousness and, 26, 29, 138
dualism and, 161–162
intrinsic existence and, 2, 73, 175n2, 191n2
malicious deceiver and, 2
mind–brain problem and, 138
pineal gland and, 48
vivisection and, 180n2
Disorders of consciousness, 23, 94–95, 193n6
Dissociative identity disorder, 113
Diving Bell and the Butterfly, The (Bauby), 97
DMT, 46, 115
Doors of Perception, The (Huxley), 47
Dreams, 156, 184n8
brain waves and, 45f
distinct states of consciousness and, 44
electroencephalography (EEG) and, 185n11
experience and, 154
measurement and, 102, 185n11
prefrontal cortex and, 63
REM sleep and, 46, 102, 185n11
Drugs
experience and, 17–18, 23, 46–47, 71, 102, 117, 162, 179n9
hallucinogens, 10, 46, 102
psychedelic, 46, 115
Dualism, 48, 161–162, 185n15, 208n15

Ecstatic experiences, 199n18
Ego, 7, 12, 18, 114–115
Eleatic Stranger, 81
Electrical brain stimulation (EBS), 64–67, 189nn22–24
Electroencephalography (EEG)
activated, 98
alpha waves and, 44
anesthesia and, 99, 101–102, 194n12
brain waves and, 44–46, 179n17
capturing dynamics and, 97
coma and, 99, 101
dissociative identity disorder and, 113
dreams and, 185n11
gamma synchrony and, 194n13
global neuronal workspace theory and, 204n16
high-density cap and, 100, 199n23
infants and, 201n14
integrated information theory (IIT) and, 100–101
neocortex and, 44

Electroencephalography (EEG) (cont.)
neural correlates of consciousness (NCC) and, 99
perturbational complexity index (PCI) and, 100f, 101–104, 196n20
pure consciousness and, 113, 116
sleep and, 45f, 46, 98, 101–102, 165, 184n9, 185n11, 199nn20,23
spatiotemporal structure of, 100
voltage fluctuations and, 44, 98, 194n12
wakefulness and, 46, 99–102, 184n9, 185n11, 199n20
zap-and-zip technique and, 100–104
Empathy, 172–173, 209n5
Encephalization quotient, 181n7
Epilepsy
brain and, 40, 50, 61, 63, 65, 99, 185n19, 188n15, 189n24, 197n5, 204n16
experience and, 61, 53, 65
Epiphenomenalism, 120–121
Ethical consequences, 169–173, 209nn1,3
Executive summary hypothesis, 124
Ex Machina (film), 129
Expander graphs, 152–154, 206n19
Experience. *See also* Consciousness
attention and, 10, 13, 21, 33, 63, 204n16
awareness and, 1, 7, 18, 23–24, 60, 158, 188n19
Causal power and, 10–11, 141–154, 158–160, 166–167, 178n1, 187n10, 205n12, 206n15, 209n3
cerebellum and, 55–58, 63, 67, 158, 162, 186nn4,6, 207n3
color and, 5–8, 14, 19, 43, 46, 48–49, 60–61, 64, 187nn8,9
computers and, 38, 141–154
defining consciousness as, 1–2, 5–6
denial of, 3–4, 161
dreams and, 44–46, 63, 88, 102, 128, 130, 154, 156, 166, 178n2, 184n8, 185n11
ecstatic, 199n18
electrical brain stimulation (EBS) and, 64–67, 189nn22–24
feeling and, 1, 5–6, 7, 12, 18, 20, 53, 60–61, 64, 141, 163, 173, 182n4, 183n3, 189n22
first-person perspective and, 11, 17, 22, 182n13
group mind and, 111, 163–167
hard problem and, 73, 190nn2,3
imitation game and, 150–151, 206n16
importance of, 169–173
integrated information theory (IIT) and, 76–77, 141–148, 151–155, 158–167, 205n12, 208nn11,13,17,20,22
mystical, 7, 114–115, 198n17, 199n18
neurons and, 7, 53–61, 64, 66–69, 150–151, 154, 158, 162–166, 186nn3,4, 187n7, 188n19, 189n25, 190nn27,30, 205n4, 206n19
pain and, 4–5, 9, 14, 23, 53, 62, 93–94, 96, 120, 156, 159, 161, 170–171, 175n5, 176n11
perception and, 5–6, 10–11, 14–21, 60–61, 161, 177n18, 178nn3,6, 179n9, 187n8, 188n19, 189n22, 190n28
phenomenology and, 2, 4, 8, 14, 17–19, 153–154, 177n12, 178n4
point of view and, 9–10, 142, 153–154, 171
posterior cortex and, 65–66
prefrontal cortex and, 61–64
psychology and, 141, 159, 161, 163, 177nn12,18, 178nn3,4, 179n9, 204n2
pure, 7, 21, 75, 105, 114–117, 121, 162, 199n20
seeing and, 5, 8–9, 11, 14, 61, 65f, 170, 177nn12,16

sentience and, 2, 157, 159, 166, 171
structure of, 6–8
subjective, 1, 3–5, 11, 14, 17–18, 22, 42–43, 57, 59, 76, 116, 119, 130, 133, 155, 177nn17,18, 179n8, 190n28
thought and, 18, 20–21, 31, 33–34, 38, 43, 46, 94, 97, 154, 166, 175n5, 185n11, 189n24
time and, 9–10
tree of life and, 31, 155–158, 169, 173, 206n2
ubiquitousness of, 155–167
visual cortex and, 60, 64–65, 206n22
wakefulness and, 6, 23, 53–55, 62, 154, 162, 165, 186n2
zombies and, 1, 5–6, 20, 29, 57, 73, 113, 119, 151, 171
Experiential responses, 7, 65–66, 113, 162, 170–171, 188n19, 189n23
Eye movements
automatisms and, 179n17
fusiform face area (FFA) and, 50
lower-level functions and, 187n13
rapid (REM), 46, 55, 102, 184n9, 185n11, 195n14
vegetative state (VS) and, 96
zombie agents and, 19–20

Faces
agnosia and, 51, 60, 187n8
blindness to, 51, 60, 186n20
brain and, 48f, 49–51, 53
confidence in perception of, 17–18
electrical brain stimulation (EBS) and, 64
electroencephalography (EEG) and, 14
neural correlates of consciousness (NCC) and, 61
perception of, 2, 5–6, 14–18, 48f, 49–51, 124, 150, 157, 189n22
Fechner, Gustav, 161
Feedback
cerebellum and, 55, 58
computation and, 135, 142, 145, 204nn1,2
recurrent or feedback processing and, 135, 142, 204nn1,2, 205nn3,4
Feedforward circuits, 58, 135, 141–145, 150, 204nn1,2
Feeling. *See* Consciousness; Experience
Flotation tank, 115
Freud, Sigmund, 19, 22, 34, 175n5, 182n4
Frontal lobes, 62–63, 188nn14,17
Frontoparietal areas, 139, 188n19, 195n16
Functionalism, 119–128
artificial intelligence (AI) and, 132–134
cognition and, 119–120, 124–125, 199n1
computers and, 128, 143–150
consciousness, 139–140
integrated information theory (IIT) and, 116, 121–126, 201n15
intelligence and, 119, 124–128, 201n12
Functional magnetic resonance imaging (fMRI), 49, 196n2, 198n12, 204nn15,16
Fusiform face area (FFA), 50, 81, 186n20
Fusiform gyrus, 50, 65f, 185n19

Gall, Franz Joseph, 41–42
Gamma waves, 44, 98–99, 194n13, 195n14, 199n23
Global ignition 139, 188n19, 195n16
Global workspace theory, 139–140, 188n19, 204nn16,17
Glutamate, 55
Go, 132, 151
Gödel, Kurt, 131
Great Ape Project, 173
Great Chain of Being, 172

Group mind, on the impossibility of 163–167

Hallucinations, 5f, 10, 18, 46, 64–65, 102, 185n12, 189n23
Hard problem, 73, 190nn2,3
Heart, 176n8, 180n2
  Aristotle on, 39, 183n3
  blood circulation and, 55, 120–121
  brainstem and, 53
  death and, 43
  as seat of reason, 39–40
  as seat of soul, 51
Hegemony, neuronal, 112–113
Hemodynamic response, 49
Hinduism, 7, 114
Hippocrates, 40
Hive mind. *See* Über-mind
Hodgkin-Huxley equations, 138, 203n11
Homunculus
  creativity and, 35f
  intelligence and, 15f, 36, 64
  motor skills and, 35f
  thought and, 35f
  unconscious, 35f, 36, 64, 124
Hot zone. *See also* Posterior hot zone
  experience and, 64–66, 71–72, 89–90, 117, 163–164, 187n10, 199n23, 208n18
  pure consciousness and, 117, 199n23
  theory of consciousness and, 71–72
Hubel, David, 134–135
Hypothalamus, 40, 54, 186n3
Hysteria, 113, 175n5

Illusion, 4, 16, 36–37, 175n5, 192n9
Imagination, 171, 176, 205n4
Imitation game, 150–151, 206n16
Inattentional blindness, 37, 183n7, 204n15
Infants, 22, 93, 128, 201n14
Information processing
  cognition and, 33–36
  computationalism and, 132, 134
  functionalism and, 121, 124–125
  integrated information theory (IIT) and, 116, 124–125
  pyramid of, 33–36
Insects, 26, 113, 156f, 157, 207n4
Integrated information theory (IIT), 74–77, 79–91
  Aaronson and, 152–153, 201n15
  axioms of (*see* Axioms of integrated information theory)
  attention and, 204n16
  causal power and, 74, 79–82, 88, 104, 116, 140–142, 145, 147, 153, 159–160, 166–167, 191n3, 196n21, 205n12
  cause-effect structure and, 79, 87–88, 116–117, 158, 163
  central identity of, 80f, 87–89, 106
  Chalmers and, 73–74
  computationalism and, 140, 204n16
  computers and, 101, 141–148, 151–154
  electroencephalography (EEG) and, 100–101
  evolution and, 121–124
  experience and, 76–77, 141–148, 151–155, 158–167, 205n12, 208nn11,13,17,20,22
  extrinsic experience and, 74–75
  functionalism and, 116, 121–126, 201n15
  Great Chain of Being and, 172
  group mind and, 111, 163–167
  hegemony and, 112–113
  information processing and, 116, 124–125
  intelligence-consciousness (I-C) plane and, 124–128
  intrinsic existence and, 80–82, 166

logic gates and, 82–87, 142, 144f, 148, 150, 153, 206n20
measurement and, 100–101, 104, 196n21
mind uploading and, 148–152
multiple minds and, 112–113
neural correlates of consciousness (NCC) and, 106–107
neutral monism and, 208n22
perturbational complexity index (PCI) and, 104
postulates of (*see* Postulates of integrated information theory (IIT))
pure consciousness and, 106–109, 112, 116–117, 198n13, 199n23
python code for computing, 91, 191n7, 193n14
quantum mechanics and, 77, 208n11
Searle and, 205n12, 208n17
sentience and, 74, 76, 166
theory of consciousness and, 74–79, 190n3, 191n7
thought and, 163, 208n13
Tononi and, 100
über-mind, 109–111, 166
Whole and, 79–81, 87–88, 106–107, 191nn3,4
Intelligence
across species, 25, 31, 171, 181n7
artificial (AI), 13, 132–134, 139, 141, 167, 191n1
brain-as-computer metaphor and, 130, 134–137
brain damage and, 187n13
brain size and, 125, 127, 200n10
computation and, 129, 132–134, 139, 176n9, 209n5
creativity and, 35f, 36, 64
experience and, 5, 33, 141–142, 151, 152f
functionalism and, 119, 124–128, 201n12
homunculus and, 15f, 36, 64
imitation game and, 150–151, 206n16
integrated information and, 158
intelligence-consciousness (I-C) plane and, 124–128, 154
IQ and, 105, 125, 151, 188n15, 200n9, 201n12
language and, 13, 38
measurement and, 200n9, 201n12
prefrontal cortex and, 64
superintelligence and, 5, 129, 209n5
Turing and, 150–151, 206n16
Internet, 11, 138
Interoceptive perception, 34
Intrinsic existence
integrated information theory (IIT) and, 80–82, 166
theory of consciousness and, 75–76
Whole and, 79–82, 191n2
Introspection, 7, 22, 31, 34, 36, 62, 177n12, 198n15
IQ, 105, 125, 151, 188n15, 200n9, 201n12
Isolation tank, 115

Jackendoff, Ray, 34
James, William, 161, 163
Jaynes, Julian, 181n11
Judgment, 15, 18, 187n13, 188n19

Kasparov, Garry, 151
Ketamine, 102, 185n12
Kundera, Milan, 173

Language
aphasia and, 29–30, 61
consciousness and, 28–31, 105, 181n11, 182n12
experience and, 31, 38, 66, 149, 169, 177n13, 178n2
Jackendoff and, 34
thought and, 28, 38
Wada test and, 196n2

Leibniz, Gottfried Wilhelm, 71–73, 82, 130–131, 160–161, 190n1, 202n3
Lévi-Strauss, Claude, 130
Lilac chaser, illusion, 16–17
Lobectomy, 62–63
Locked-in syndrome (LIS), 95, 97, 102–103
Logic gates, 82–87, 142–150, 153, 206n20
LSD, 46, 185n12

Mach, Ernst, 4
Machine learning, 103, 129, 132, 135, 144, 184n10
Magnetoencephalography (MEG), 185n12, 194n13, 204n16
Main complex of consciousness, 87
Markram, Henry, 203n12
Marr, David, 16
Materialism, 72, 161, 175n5
Maximally irreducible cause-effect structure, 80, 87–89, 164
  inactive cortex and, 116–117
  integrated information theory (IIT) and, 79, 87–88, 116–117, 158, 163
  Whole and, 79, 84–90, 192n10, 193n14
Maxwell equations, 125, 190n30,
McGinn, Colin, 71
Mechanism, 69, 72, 74–75, 82–90
Mereology, 192n9
Mescaline, 46
Metacognition, 63, 188n18
Metzinger, Thomas, 177n16
Mill thought experiment (Leibniz), 71–72
Mind-as-software mythos, 129–130, 134f
Mind blanking, 21–22, 37, 179n14
Mind blending. *See* Über-mind
Mind blindness, 59–61
Mind–body problem, xi, 11, 49, 104, 120, 164f
Mind uploading, 148–152
Mind wandering, 113, 115
Minimally conscious state (MCS), 97, 102–103
Moore's law, 131
Morgan, C. L., 182n2
Motion blindness, 61, 187n8
Motor skills
  acquiring new, 21
  brainstem and, 53
  cerebellar damage and, 186n5
  computationalism and, 133, 140
  dreams and, 44
  drugs and, 46
  experience and, 5, 9, 20–21, 53, 57–66, 154
  feedforward networks and, 58, 135, 141–145, 150, 204nn1,2, 205nn3,4
  functionalism and, 120, 122, 124–125, 128, 199n2
  homunculus and, 35f
  reflexes and, 20 (*see also* Reflexes)
  selective attention and, 38
  sensorimotor tasks and, 9, 38, 44, 113
  supplementary motor cortex and, 193n7
  visuomotor behavior and, 5, 133, 199n2, 204n2
  winking and, 20, 185n19
  zombie agents and, 20–21
Multiple minds in a brain, 112–113
Multiple realizability, 193n12
Multiverse, 11, 77, 178n1, 191n6
Mystical experiences, 7, 114–115, 198n17, 199n18

Neocortex
  animal consciousness and, 27, 157–158, 180n6
  electroencephalography (EEG) and, 44
  mind blindness and, 59–61
  neural correlates of consciousness (NCC) and, 61

Neural correlates of consciousness (NCC)
 blood flow and, 49
 brain and, 47–51
 brainstem and, 53–55
 cerebellum and, 67
 computationalism and, 139–140
 correlative evidence and, 63–64
 cortices and, 61, 63–64, 67, 188n20
 Crick and, 47–49, 185n14
 electroencephalography (EEG) and, 99
 faces and, 51, 61
 fusiform face area (FFA) and, 50
 global workspace theory and, 139–140
 hot zones and, 64–66, 71–72, 89–90, 117, 163–164, 187n10, 199n23, 208n18
 integrated information theory (IIT) and, 106–107
 ontologically neutral, 48–49
 rigorous definition of, 185n14
 sleep and, 51
 theory of consciousness and, 71, 73
 Whole and, 106–107
Neural networks, 79, 122–123, 132, 135, 141, 206n19
Neuromorphic electronic hardware, 150
Neuron doctrine, 42
Neurons
 action potentials and, 42, 58, 75, 98, 117, 138, 203n11
 coining of term, 184n6
 computationalism and, 134–140, 202n8, 203nn11,12, 204nn16,17
 crown of thorns, 198n9
 experience and, 7, 53–61, 64, 66–69, 150–151, 154, 158, 162–166, 186nn3,4, 187n7, 188n19, 189n25, 190nn27,30, 205n4, 206n19
 footprints of consciousness and, 39
 functionalism and, 125–128, 200n10, 201nn12,14,15
 integrated information theory (IIT) and, 76
 number of, 185n16, 180nn5,6, 181n7
 pyramidal, 58–61, 117, 136, 154, 187n7, 190n27
 Ramón y Cajal and, 42
 simple cells and, 135
 theory of consciousness and, 71, 75–76, 190n3, 192n12
 Whole and, 81–82, 88–90
New Testament, 40, 71, 173
Nietzsche, Friedrich, 22, 129, 177n16
Nonhuman Rights Project, 173
Noradrenaline, 55, 107

Objective measures of perception, 14
Occipital-parietal cortex, 61, 64
On the Sacred Disease (treatise), 40
Organoids. *See* Cerebral organoids

Pain
 anesthesia and, 47, 62, 93–94
 animal, 26–29, 209n4
 brain and, 43, 47–48, 50
 chronic, 194n10
 experience and, 4–5, 9, 14, 23, 53, 62, 93–94, 96, 120, 156, 159, 161, 170–171, 175n5, 176n11
Panpsychism, 160–163, 166, 192n7, 208n13
Penfield, Wilder, 65–66, 189n23
Penrose, Roger, 68, 190n29
Perception
 as active process, 16
 agnosia and, 51, 60, 187n8
 brain and, 44–47, 50
 color, 187n8
 computation and, 134, 139
 drugs and, 46–47
 faces and, 1–2, 5–6, 14–18, 48f, 49–51, 124, 150, 157, 189n22
 judgment and, 15, 18
 limits of, 183n7

Perception (cont.)
measurement and, 14, 94, 195n16
motion, 60, 187n8, 189n22
psychoanalysis and, 182n4
reliability of, 17
subjective confidence in, 10, 17–18, 22, 63, 93, 179n8, 188n19
tactile, 188n19
unconscious priming and, 18, 179n9
Perturbational complexity index (PCI), 100f, 101–104, 196n20
Phenomenal consciousness, 19, 179n11. *See also* Consciousness
Phenomenology, 2, 4, 8, 14, 17–19, 153–154, 177n12, 178n4
axioms of, 75
functionalism and, 201n15
pure consciousness and, 115–116
theory of consciousness and, 74–76
Phosphenes, 64, 189n22, 206n22
Phrenology, 41–42
Physicalism, 48–49, 72, 161, 199n24, 208nn15,21,25
Physical substrate of consciousness, 87
Plato, 39, 81, 138, 161, 191n3
Point of view
experience and, 9–10, 142, 153–154, 171
Whole and, 80–81, 86, 89
Posterior cortex, 162
brain injuries and, 187n10
electrical brain stimulation and, 66
hot zone of, 64–66, 71–72, 89–90, 117, 163–164, 187n10, 199n23, 208n18
imaging techniques and, 189n21
neural correlates of consciousness (NCC) and, 61
posterior temporal cortex, 189n21
sensory experience and, 65–66
visual awareness negativity and, 195n16
Whole and, 163–164
Posterior hot zone. *See* Hot zone
Postulates of integrated information theory (IIT), 75–76
composition postulate, 82–84
exclusion postulate, 86–87, 90, 107, 109, 112, 148, 163, 165–166, 192n7
information postulate, 84–85
integration postulate, 85–86
intrinsic existence postulate, 80–81
Prefrontal cortex, 61–64
Prosopagnosia, 51
Psilocybin, 46, 185n12
Psychedelic drugs, 46, 115
Psychophysics, 12, 14, 17–18, 161, 178n4
P3b, 99, 195n16
Pure consciousness. *See* Pure experience
Pure experience, 7, 75, 114–117, 162, 199n20
Purkinje cells, 56–58
Pyramidal neurons, 58–61, 117, 136, 154, 187n7, 190n27

Quantum mechanics
brain and, 68–69, 190nn28–30
consciousness, 68–69, 190n28
experience and, 151–152, 161, 190nn28–30
integrated information theory (IIT) and, 208n11
theory of consciousness and, 73, 77

Ramón y Cajal, Santiago, 42
Rapid eye movement (REM), 46, 55, 102, 184n9, 185n11, 195n14
Recurrent processing, 142
Reentry processing, 142
Reflexes
adaptive, 20
behaviors resembling, 20, 23
blinking, 20, 95, 97
classical, 20
comatose patients and, 23
Lazarus, 194n8

lower-level functions and, 187n13
spinal cord and, 20
zombie agents and, 20
Retina
computation and, 133–134, 136, 202nn4,8,10
vision and, 15f, 16, 33, 58, 64, 68, 133–134, 136, 202nn4,8,10
Ruby, 6, 28, 163
Russell, Bertrand, 161

Saint Augustine, 2, 73, 80, 175n3
Schiavo, Terri, 23, 179n17
Schrödinger, Erwin, xi, 68, 161
Searle, John, 4, 205n12, 208n17
Seeing
attentional lapses and, 183n7 (*see also* Mind blindness)
blinking and, 20, 95, 97
brain and, 48, 50, 183n3
experience and, 5, 8–9, 11, 14, 61, 65f, 170, 177n12, 177n16
Seizures, 30, 40, 61, 63, 65, 105, 187n11, 189n23, 195n14
Selective attention. *See* Attention
Self-consciousness. *See* Consciousness, self
Sensorimotor experience, 9, 38, 44, 113
Sensory-cognitive-motor actions, 20–21, 120
Sentience
animal consciousness and, 25, 29, 157, 159, 170–172
Buddhism and, 173
experience and, 2, 157, 159, 166, 171
integrated information theory (IIT) and, 74, 166
theory of consciousness and, 74, 76
Serotonin, 46, 55
Seung, Sebastian, 151
Sex, 8, 18, 37, 63, 114, 162
Shannon, Claude, 84–85, 192n6, 205n12
Shannon information, vs. integrated information, 84–85, 192n6, 205n12
Silesius, Angelus (Meister Eckhart), 7, 114
Silicon Valley, xiv, 4, 129
Simple cells, 135
Singer, Peter, 172
Sleep
anesthesia and, 29, 47, 62, 93–94, 99, 101–102, 106, 128, 194n12, 195n14
brain waves and, 23, 46
conscious states and, 44
consumer devices for, 46, 184n10
creativity and, 36
deep, 34, 45f, 46, 48f, 51, 101–102, 114, 116, 165, 186n3, 195n17, 199nn20,21
dreams and, 44–46, 63, 88, 102, 128, 130, 154, 156, 166, 178n2, 184n8, 185n11
electroencephalography (EEG) and, 45f, 46, 98, 101–102, 165, 184n9, 185n11, 199nn20,23
integrated information and, 162
neural correlates of consciousness (NCC) and, 51
quiet restfulness and, 195n17
REM and, 46, 55, 102, 184n9, 185n11, 195n14
slow wave sleep (*see* Sleep, deep)
vegetative state (VS) and, 47, 96, 102–103, 171, 193n6
Sleeping sickness (*encephalitis lethargica*), 54
Sleepwalkers, 23
Snaprud, Per, 104
Socrates, 13, 39, 161
Solipsism, 12, 155
Soul
animals and, 25
anima mundi or the world soul, 166
sensorium commune and, 40
Cartesian, 26, 29

Soul (cont.)
essence of, 7
immortal, 25
panpsychism and, 160–161
Soul 2.0, xiv
Whole and, 192n9
Sperry, Roger, 106, 197n4
Spinal cord
brainstem and, 53, 58
reflexes and, 20
trauma to, 53
Split-brain patients. *See* Brain, split
Spurzheim, Johann, 41–42
Star Trek, 94, 111, 165
Stem cells, 126–128, 154, 201n14
Strawson, Galen, 4, 176n7
Strokes, 23, 29, 47, 56, 60, 113, 186n20
Subjective measures, 17–18, 22, 179n8
Suicide, 152, 194n10
Supercomputers, 131, 138, 149–152
Supplementary motor cortex, 193n7
Synapses
axons and, 42
brain and, 178n6, 203n12
chemical, 184n7
computation and, 135, 137
dendrites and, 42, 137, 190n27
electrical, 184n7
hot zone and, 72
neuron doctrine and, 42
proteins and, 196n21

Tegmark, Max, 191n6
Temperature, 34, 53, 69, 115, 159–160, 190n30
Temporo-parietal-occiptal cortex, 61, 189n22
Thalamus, 54, 58, 64, 112, 186nn3,4
Theory of consciousness
causal power and, 74–75
feeling and, 73–76
global workspace theory and, 139–140
integrated information theory (IIT) and, 74–79, 190n3, 191n7
intrinsic existence and, 75–76
materialism and, 72, 161, 175n5
mill thought experiment and, 71–72
neural correlates of consciousness and, 71, 73
neurons and, 71, 75–76, 190n3, 192n12
phenomenology and, 74–76
physicalism and, 72
quantum mechanics and, 73, 77
sentience and, 74, 76
Third-person properties, 11, 14, 17, 22
Thought
dreams and, 185n11 (*see also* Dreams)
electrical brain stimulation and, 189n24
experience and, 18, 20–21, 31, 33–34, 38, 43, 46, 94, 97, 154, 166, 175n5, 185n11, 189n24
homunculus and, 35f
integrated information theory (IIT) and, 163, 208n13
Jackendoff on, 34
language and, 28, 38
suicidal, 194n10
Thought experiments, 68, 71–72, 111, 117
Tip-of-the tongue phenomenon, 34–35
Tononi, Giulio, 74, 100, 144, 153–154, 177n19, 191n7, 205n12
Tree of life, 31, 155–158, 169, 173, 206n2
Tristan und Isolde (Wagner), 111
Turing, Alan, 131–134, 137t, 147, 150–151, 202n3, 206n16
Twins with conjoint skulls and brains, 109, 197n7

Über-mind, 109–111, 166
Uncertainty principle, 68
Unconscious homunculus, 35f, 36, 64, 124

Unconscious priming, 18, 179n9
Universal computation, 130–131

Vegetative state (VS), 47, 96, 102–103, 171, 193n6
Virtual reality, 17
Vision
computation and, 19, 132, 205n4
electroencephalography (EEG) timing and, 14
failures of, 37
hallucinations and, 5f, 10, 18, 46, 64–65, 102, 185n12, 189n23
mind blindness and, 59–61
photons and, 1, 33, 68, 136, 202n4
retina and, 1, 15f, 16, 33, 58, 64, 68, 133–134, 136, 202nn4,8,10
swiftness of, 14
winking and, 20, 185n19
Visual awareness negativity (VAN), 195n16
Visual cortex
computation and, 134–135
experience and, 60, 64–65, 206n22
gamma-range synchronization and, 194n13
phosphenes and, 189n22
Visuomotor behavior, 5, 133, 199n2, 204n2
von Economo, Baron Constantin, 54
von Neumann, John, 133, 145, 150, 171, 190n28
von Waldeyer-Hartz, Wilhelm, 184n6

Wakefulness
under anesthesia, 93
brain and, 43–44, 46, 51
electroencephalography (EEG) and, 46, 99–102, 184n9, 185n11, 199n20
experience and, 6, 23, 53–55, 62, 154, 162, 165, 186n2
measurement and, 93, 96, 99–104
vegetative state (VS) and, 47, 96, 102–103, 171, 193n6
Westworld (TV series), 129
Whitehead, Alfred North, 161, 176n7
Whole
causal power and, 79–89, 191nn1,3, 192nn9,12
cause-effect structure and, 79, 84–90, 192n10, 193n14
central identity of IIT and, 80f, 87–89, 106
color and, 192n10
definition of, 87, 192n9
group mind and, 111, 163–167
integrated information theory (IIT) and, 79–81, 87–88, 106–107, 191nn3,4
intrinsic existence and, 79–82, 191n2
logic gates and, 82–87, 142, 144f, 148, 150, 153, 206n20
mereology and, 192n9
neural correlates of consciousness (NCC) and, 106–107
physical substrate of consciousness, 87
posterior cortex and, 163–164
Whole brain emulation, 137–139
Wiesel, Torsten, 134–135
Willis, Thomas, 41, 184n4
Winks, 20, 185n19
Woolf, Virginia, 21
Wundt, Wilhelm, 161

Zap-and-zip technique, 100–104
Zombies
abductive reasoning and, 12
Chalmers on, 73–74, 94
hard problem and, 73, 190nn2,3
lack of experience and, 1, 6, 20, 29, 57, 73, 113, 119, 151, 171
nonconscious agents and, 19–22
unconscious behaviors and, 119–120